Groundwater Management for Sustainable Agriculture in the North China Plain

This book is a unique text that explores recent research on the management of sustainable groundwater use in the North China Plain (NCP), where aquifers are suffering the most severe over-pumping in the world and have caused serious ecological degradation. It contains research conducted by the editor and his teams on several projects over the past 18 years.

Key topics covered include:

- comprehensive scheme and pathways to achieve sustainable groundwater management
- description of theoretical basis for water saving and technologies developed in practice at field scale
- adjusting cropping patterns and planting structure to reduce the cultivation intensity to a suitable extent
- soft measures such as water metering, pricing, and water marketing being applied in groundwater management practice in the NCP.

It will be an invaluable resource to graduate students, education and research staff, and agriculture or water resources authorities.

Groundwater Management for Sustainable Agriculture in the North China Plain

Edited by Yanjun Shen

CRC Press
Taylor & Francis Group
Boca Raton London New York

CRC Press is an imprint of the
Taylor & Francis Group, an **informa** business

Designed cover image: Hongjun Li

First published 2025
by CRC Press/Balkema
4 Park Square, Milton Park, Abingdon, Oxon, OX14 4RN

and by CRC Press/Balkema
2385 NW Executive Center Drive, Suite 320, Boca Raton FL 33431

CRC Press/Balkema is an imprint of the Taylor & Francis Group, an informa business

© 2025 selection and editorial matter, Yanjun Shen; individual chapters, the contributors

The right of Yanjun Shen to be identified as the author of the editorial material, and of the authors for their individual chapters, has been asserted in accordance with sections 77 and 78 of the Copyright, Designs and Patents Act 1988.

All rights reserved. No part of this book may be reprinted or reproduced or utilised in any form or by any electronic, mechanical, or other means, now known or hereafter invented, including photocopying and recording, or in any information storage or retrieval system, without permission in writing from the publishers.

Although all care is taken to ensure integrity and the quality of this publication and the information herein, no responsibility is assumed by the publishers nor the author for any damage to the property or persons as a result of operation or use of this publication and/or the information contained herein.

British Library Cataloguing-in-Publication Data
A catalogue record for this book is available from the British Library

Library of Congress Cataloging-in-Publication Data
Names: Shen, Yanjun, editor.
Title: Groundwater management for sustainable agriculture in the North China Plain / edited by Yanjun Shen
Description: First edition | Boca Raton, FL : CRC Press/Balkema, 2025 | Includes bibliographical references and index
Identifiers: LCCN 2024011559 (print) | LCCN 2024011560 (ebook) | ISBN 9781032116747 (hardback) | ISBN 9781032116754 (paperback) | ISBN 9781003221005 (ebook)
Subjects: LCSH: Water-supply, Agricultural—China—North China Plain. | Groundwater—China—North China Plain—Management. | Water in agriculture—China—North China Plain.
Classification: LCC S494.5.W3 G76 2025 (print) | LCC S494.5.W3 (ebook) | DDC 333.91/30951—dc23/eng/20240617
LC record available at https://lccn.loc.gov/2024011559
LC ebook record available at https://lccn.loc.gov/2024011560

ISBN: 9781032116747 (hbk)
ISBN: 9781032116754 (pbk)
ISBN: 9781003221005 (ebk)

DOI: 10.1201/9781003221005

Typeset in Times New Roman
by codeMantra

Contents

Preface	*vii*
Acknowledgements	*ix*
About the editor	*x*
List of contributors	*xi*
List of abbreviations	*xii*

PART 1
Groundwater depletion and sustainable use issue 1

1 **Population growth, food demand and water crisis in the North China Plain** 3
   YANJUN SHEN, HONGWEI PEI, AND DENGPAN XIAO

2 **Integrated paths to sustainable groundwater management** 17
   YANJUN SHEN AND DENGPAN XIAO

PART 2
Water-saving agriculture: One drop more crops 29

3 **Water-saving agriculture: Principles, regulation, and framework of practices** 31
   HUIXIAO WANG, XIAOHONG REN, YANJUN SHEN, YONGQING QI, AND CHANGMING LIU

4 **Water budgets, ET partitioning, and implications to water saving** 50
   YUCUI ZHANG, YANJUN SHEN, AND FAN LIU

5 **Improving water productivity by integrating bio-, agro- and engineering measures** 71
   XIYING ZHANG

PART 3
Water adaptive agriculture: Reducing pumping intensity   95

6 Agricultural land use change and impacts on groundwater   97
   YANJUN SHEN AND YUCUI ZHANG

7 Adjustment of field cropping patterns:
   A better intensity   112
   YONGQING QI, JIANMEI LUO, YANJUN SHEN, DENGPAN XIAO,
   AND SUYING CHEN

8 Optimization of regional cropping structure:
   A better planning   127
   JIANMEI LUO, YANJUN SHEN, AND YONGQING QI

PART 4
Better water management for sustainable agriculture   145

9 Managing groundwater quality and quantity
   of irrigated farmland   147
   LEILEI MIN, MEIYING LIU, YONGQING QI, AND YANJUN SHEN

10 Utilization of brackish and other abnormal water in irrigation   161
   YUCUI ZHANG, YANJUN SHEN, AND YONGQING QI

11 Innovation of policy system and smart irrigation technologies for a better
   groundwater governance   173
   HONGJUN LI, YONGQING QI, TAO QUAN, YANJUN SHEN, AND DENGPAN XIAO

   Index   *193*

Preface

Groundwater serves as an important and easy way to acquire water resources, sustaining human society and natural ecosystems on the Earth. Under the growing pressure of food demand and rapid economic development, groundwater over-exploitation occurs in arid and semi-arid areas, such as the US High Plains, Indus Plain, and northern China. At present, groundwater over-exploitation has become a widespread issue all over the world. The regions with severe groundwater depletion are facing a series of environmental consequences such as drying up of rivers, shrinkage and disappearance of wetlands, and land subsidence.

Groundwater sustainability is crucial to agricultural and socioeconomic development, especially in the North China Plain (NCP), which is considered as the "Grain Basket" of China and has a history of more than 50 years of groundwater over-exploitation. Hence, it is vital to wisely manage groundwater resources to sustain the increasing agricultural production demand, booming population, and rapid growth in economy.

We conducted a series of systematic research focused on agricultural water conservation and sustainable groundwater management in the NCP in the past two decades, and gained a wealth of knowledge and hands-on experience in water management and its sustainability. We summarized our researches, experiences, and techniques from our previous work and compiled this book, which we hope to introduce to our international peers and community, who are interested in this topic, especially those still suffering from severe water shortage issues.

The book consists of 11 chapters, divided into four parts. Part 1, Chapters 1 and 2, demonstrates the background and facts of groundwater over-exploitation in the NCP, along with the framework for sustainable management pathways, outlining the history of water crisis in the NCP and laying the groundwork for possible solutions. Part 2 incudes Chapters 3–5, which clarifies the theoretical basis for water-saving practices and the technologies developed in practice, while exploring the water-saving potential at the field scale. Part 3, including Chapters 6–8, outlines an improved way of adjusting cropping patterns and planting structure to mitigate the intensity of agriculture in order to align agricultural productivity with water resources carrying capacity. Part 4, Chapters 9–11, shows the efforts made to achieve sustainable use of water resources from the perspective of management of various aspects, such as managing the groundwater quality and quantity; increasing water sources of brackish water, rainwater, reclaimed water for irrigation use; innovation in water policies and the supporting intelligent technologies.

This book starts with the measures of improving the water use efficiency, and aims at preventing groundwater over-exploitation and alleviating the pressure of groundwater utilization. From theories, to schemes, and finally to practices, this book includes some works of our water-saving research groups and about 12 master and doctoral students' theses at College of Advanced Agricultural Sciences, the University of the Chinese Academy of Sciences (UCAS) and the Center for Agricultural Resources Research, Institute of Genetics and Developmental Biology, the Chinese Academy of Sciences. The book reflects our attempts at conducting studies to carry out agricultural water conservation and groundwater protection in the high-intensity agricultural production area in China, so as to make coordinated development of groundwater utilization and food production.

Due to the limitations of our knowledge and time, and also the language ability, there could be some deficiencies in this book. Please feel free to contact us if you find something inappropriate or incorrect.

Acknowledgements

The editor is grateful to the Natural Science Foundation of Hebei Province (D2021503001), National Scientific Foundation of China (NSFC, 41930865, 41471027, 41877169), the Ministry of Science and Technology (2016YFC0401403, 2023YFD1900801), and the Project for Innovative Capacity Improvement in Hebei Province (225A4201D) for their support of the research projects. Many thanks to our professional colleagues for their invaluable efforts in editing and compiling the chapters and ensuring that each chapter is at its best. I appreciate Dr. Hanbing Jiang, Mr. Mengzhu Liu, and Ms. Di Geng for their great work in helping finalize the editing of the drafts.

About the editor

Yanjun Shen is currently a professor at the Center for Agricultural Resources Research, Institute of Genetics and Developmental Biology, Chinese Academy of Sciences, he is also an adjunct professor at College of Advanced Agricultural Sciences, the University of the Chinese Academy of Sciences. He is studying agricultural hydrology, mainly working on the interdisciplinary field of eco-hydrology, agronomy, and groundwater sciences. He also pays close attention to the impacts of agricultural activities on hydrology and ecosystems at basin or regional scales.

Contributors

Suying Chen Center for Agricultural Resources Research, Institute of Genetics and Developmental Biology, Chinese Academy of Sciences, Shijiazhuang, China

Hongjun Li Center for Agricultural Resources Research, Institute of Genetics and Developmental Biology, Chinese Academy of Sciences, Shijiazhuang, China

Changming Liu Institution of Geographic Science and Natural Resources Research, Chinese Academy of Sciences, Beijing, China

Fan Liu Center for Agricultural Resources Research, Institute of Genetics and Developmental Biology, Chinese Academy of Sciences, Shijiazhuang, China

Meiying Liu Center for Agricultural Resources Research, Institute of Genetics and Developmental Biology, Chinese Academy of Sciences, Shijiazhuang, China

Jianmei Luo School of Land Science and Space Planning, Hebei GEO University, Shijiazhuang, China

Leilei Min Center for Agricultural Resources Research, Institute of Genetics and Developmental Biology, Chinese Academy of Sciences, Shijiazhuang, China

Hongwei Pei Hebei University of Architecture, Zhangjiakou, China

Yongqing Qi Center for Agricultural Resources Research, Institute of Genetics and Developmental Biology, Chinese Academy of Sciences, Shijiazhuang, China

Tao Quan Center for Agricultural Resources Research, Institute of Genetics and Developmental Biology, Chinese Academy of Sciences, Shijiazhuang, China

Xiaohong Ren College of Water Sciences, Beijing Normal University, Beijing, China

Huixiao Wang College of Water Sciences, Beijing Normal University, Beijing, China

Dengpan Xiao College of Geography Science, Hebei Normal University, Shijiazhuang, China

Xiying Zhang Center for Agricultural Resources Research, Institute of Genetics and Developmental Biology, Chinese Academy of Sciences, Shijiazhuang, China

Yucui Zhang Center for Agricultural Resources Research, Institute of Genetics and Developmental Biology, Chinese Academy of Sciences, Shijiazhuang, China

Abbreviations

BAU	business as usual scenario
BI	basin irrigation
BTH	Beijing-Tianjin-Hebei Region
DI	drip irrigation
DTW	depth to groundwater level
E	evaporation
EM	early sown maize
ET	evapotranspiration
FAO	Food and Agriculture Organization of the United Nations
FI	flood irrigation
G	soil heat flux
GCO	grain-cotton-oil pattern
GCP	grain-cotton-sweet potato pattern
GO	grain-oil pattern
GSPAC	Groundwater-Soil-Plant-Atmosphere Continuum
H	sensible heat flux
HI	harvest index
IoT	Internet of Things
LAI	leaf area index
LE	latent heat flux
MODIS	moderate resolution imaging spectroradiometer
N-HBP	Northern Hebei Plain
N-HNP	Northern Henan Plain
NDVI	normalized difference vegetation index
NEE	net ecosystem carbon exchange
NW-SDP	Northwestern Shandong Plain
RLD	root length density
S-HBP	Southern Hebei Plain
SDI	surface drip irrigation
SI	tube-sprinkler irrigation
SNWDP	South-to-North Water Diversion Project
SPAC	Soil-Plant-Atmosphere Continuum
SSDI	subsurface drip irrigation

SSF	self-sufficiency in the main agricultural products
T	transpiration
WD	water deficit
WP	water productivity
WUA	water user authority
WUE	water use efficiency

Part 1

Groundwater depletion and sustainable use issue

Chapter 1

Population growth, food demand and water crisis in the North China Plain

Yanjun Shen, Hongwei Pei, and Dengpan Xiao

1.1 Introduction

The world's population reached 8 billion on 15 November 2022 (United Nations Department of Economic and Social Affairs, Population Division, 2022), and is projected to increase to about 9 billion by 2050 (UNDP, 2007). As a result, feeding a growing, urbanized and affluent population in a rapidly globalized world will be a global challenge (Hanjra and Qureshi, 2010). In response to population growth and rising incomes, the world's demand for cereals and meat has been projected to increase by 65% and 56%, respectively (de Fraiture et al., 2007). Since water is a key driver of agricultural production, water scarcity remains the primary constraint on global food production. More affluent populations have tended to diversify diets towards animal food items (Popkin, 2003) which require several multiples of water per calorie of dietary energy (Molden et al., 2010). In China, meat demand (including demand for beef, pork, eggs and more dairy products) or calorie consumption has grown from less than 100 kcal/capital/day to more than 600 kcal/capital/day between 1961 and early 2003. Fulfilments of calorie and dairy requirements will translate into even higher water demand if more calories are supplied from meat (Rosegrant and Cline, 2003). Overall, an essential challenge facing agriculture in the 21st century is how to feed a world with a continuously growing and increasingly affluent population with greater calories demand.

Water for agriculture is critical for future global food security (Xiao et al., 2020). Food security should not lose sight of surging water scarcity (Hanjra and Qureshi, 2010). Water scarcity refers to a situation where there is insufficient water to satisfy normal human water needs for food, drinking and other uses, implying an excess of water demand over available supply (Falkenmark, 2007). Being the largest user of water, irrigation is the first sector to lose out as water scarcity increases (Molden, 2007) for water scarcity can cut production and adversely impact food security (Hanjra and Qureshi, 2010). Continued increase in demand for water by agricultural and non-agricultural uses has put irrigation water demand under greater scrutiny and threatened food security (Shen and Chen, 2010). Reduction in irrigation water will cause decline in food production. Irrigation has helped boost agricultural yields and outputs in arid, semi-arid and even humid environments and stabilizes food production. At the same time, the limited easily accessible freshwater resources in rivers, lakes and shallow groundwater aquifers are dwindling due to over-exploitation and water quality degradation (Tilman et al., 2002). Major food-producing areas such as the Punjabs of India and Pakistan, and the central and northern areas of China suffer from the depletion of aquifers and the reallocation of water resources from irrigation to growing urban cities, posing implications for weakening food security. The severity of the water crisis has prompted the United Nations to conclude that it is water scarcity, not the lack of arable land, that will be the major constraint to increasing food production over the next few decades (UNDP, 2007). Some of the most densely populated regions of the world, such as the Mediterranean, the Middle East, India, China and Pakistan, are predicted to face severe water shortages in the coming decades (Postel and Wolf, 2001). Further, it is believed that climate change will exacerbate water scarcity and pose more risks on agricultural production. Even if new supplies are added to existing ones, water might not be sufficient for increasing food demand, or to cope with increasing drought events.

Groundwater depletion occurs almost all over the world, even in the centre of the extremely arid region. A major reason for the widespread occurrence of groundwater depletion is that it is used as an important source of water supply (irrigation water, drinking water, domestic water, industrial water, municipal water, etc.) for most people and natural ecosystem's transpiration (Hanjra and Qureshi, 2010). Groundwater plays a key role in food production, accounting for over 40% of the global consumptive use in agricultural irrigation by 20% of global arable land (Siebert et al., 2010). Meanwhile, groundwater is also a crucial factor in the natural ecosystem, such as sustaining stream flow during dry periods, and is vital to many lakes, rivers and wetlands as well. During the past half-century, the world had witnessed a growing pressure on groundwater resources, which in many cases induced groundwater over-exploitation beyond sustainable levels and increased agriculture, industry and the urban-sourced pollution levels (Min et al., 2015). A major concern in maintaining future water supplies is the continuing overdraft of groundwater resource. Therefore, the availability of groundwater for irrigation declines (Xiao et al., 2020). Furthermore, climate change, land use change and population growth are posing a variety of threats on groundwater resources globally, thereby impacting both quantity and quality of the groundwater (Foley et al., 2011). In many agricultural regions of the world, either arid and semi-arid areas or humid and semi-humid areas, the over-exploitation of groundwater led to rapid draining of aquifers which then resulted in ecosystem degradation (Hanjra and Qureshi, 2010).

The North China Plain (NCP, Figure 1.1) (35°00′–40°30′N, 113°12′–119°50′E) is bound by Taihang Mountains to the west, Yan Mountains to the north, Bo Sea to the east and Yellow River to the south. There were 117 million people living in the NCP, and the population density in the NCP was estimated to be 818 persons/km^2 (Pei et al., 2015). With the pressure of increasing population and the progress in mechanization and electrification that happened since the 1960s, the double cropping system of wheat and corn has experienced rapid expansion throughout the NCP, especially in the piedmont areas (Pei et al., 2017). Meanwhile, surface water was interrupted by hundreds of dams in the Taihang Mountains and Yan Mountains, and almost all rivers dried up in the NCP. Groundwater accounted for ~70% of irrigation water while the remaining was from the Yellow River, and flood irrigation is the dominant method for both the surface water and groundwater irrigation (Pei et al., 2015). In the piedmont area of the NCP, nearly 100% irrigation water was from groundwater and the intensive irrigation resulted in dozens of groundwater depression cones. The NCP is a global hotspot on groundwater sustainable research and is also crucial in the Chinese food security (Yuan and Shen, 2013). The NCP shares 6.3% croplands of China (7.8 million hectares, 2007), but produces about 11% of grain of China, especially in China's supply of wheat (25%) and corn (18%) (Pei et al., 2015).

1.2 Growing population and tightening water for food in the NCP

Population is the main driving force of global water resources demand. It is reported that there were 2.53 billion people living presently in basins with high degree of water stress, and the

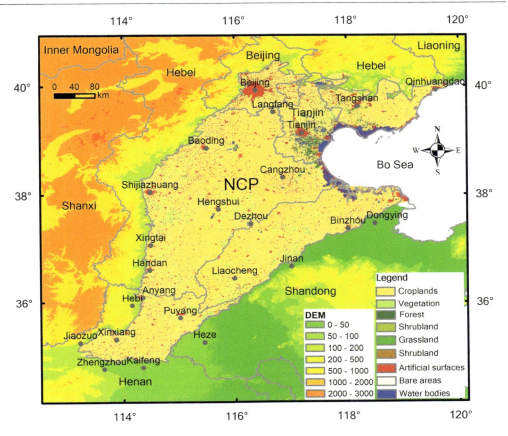

Figure 1.1 Location and land use map of the NCP in China.

number will be likely to increase to 4.16–7.58 billion in 2050 (Shen et al., 2014). The water resources per capita (WPC) of the three main provinces/municipalities (Beijing, Tianjin and Hebei, so-called BTH region hereafter) in the NCP were as low as 118 m^3/capita in Beijing in 2020 (Water Resources Bureau of Beijing, 2021), while the WPC values were a little higher in Tianjin (201 m^3/capita in 2020) (Water Resources Bureau of Tianjin, 2021) and Hebei (196 m^3/capita in 2020) (Water Resources Bureau of Hebei, 2021). The WPC value in Beijing, Tianjin and Hebei were 1/47, 1/27 and 1/28 of the world average value (5,500 m^3/capita in 2020) (The World Bank, 2023), respectively. Therefore, the shortage of water resource has already been the most serious limiting factor in the NCP, for both economic development and natural ecosystem.

Over the past 70 years, both growing population and increasing living standard in the NCP resulted in rapid rise of water resource demand. Based on the long-term population statistics datasets, the population in BTH region has increased by 169% during 1952–2019 (Figure 1.2). However, due to the differentiation of birth distribution in urban and rural areas, population growth experienced significant spatial variations. The population in Beijing rose by 347% during 1952–2019, while the growth percentage in Tianjin and Hebei was 256% and 131%,

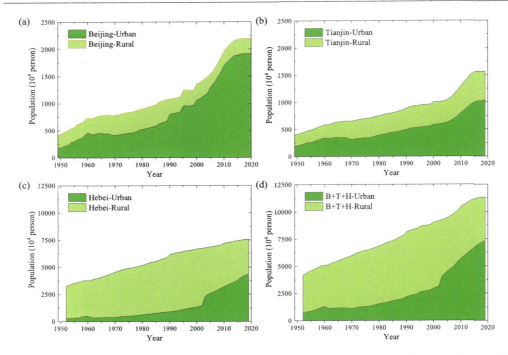

Figure 1.2 Population booming in the NCP during 1949–2020. Part labels (a), (b), (c), and (d) refer to the rural and urban population change for the Province/Municipality Beijing, Tianjin, Hebei, and their total. Due to limitation in data accessibility, the date stages for Beijing (B), Tianjin (T) and Hebei (H) were 1949–2020, 1949–2019 and 1952–2019, respectively.

respectively. Besides the total population, the population structure of urban and rural also showed variations. Beijing had the highest urban population percentage (87% in 2019), followed by Tianjin (66% in 2019) and Hebei (57% in 2019).

Mechanization and agricultural hydraulic engineering projects in the 1960s made double cropping of wheat (from October to next June) and maize (from June to October) possible throughout the NCP (Xiao et al., 2017). Irrigation water in the NCP is primarily from groundwater, accounting for ~70% in 2011, and the remaining 30% is from the Yellow River in the south, due to the cut-off of surface water by the reservoirs in the mountains. The piedmont region has the highest irrigation percentage with mean 58% of land area, while the coastal region has the lowest with mean 26% (Figure 1.3). At the provincial level, about 70% of the irrigated cropland was fed by groundwater (Pei et al., 2015), also with some scattered surface irrigation districts in the piedmont, and some large irrigation districts in the southern part of the NCP (2.1 mha, ~30%) supported by the Yellow River (Tian, 2010).

As the main part of the NCP, the yield and grain production of maize and wheat in Hebei Province have been increasing since the 1950s (Figure 1.4), with gentle decrease in only some special periods. The planted area of maize and wheat in Hebei Province has increased since the 1950s, and is more variable than the yield and production. There were three periods with significant decline for both wheat and maize planted area. The first period of continuous

Population growth, food demand and water crisis 7

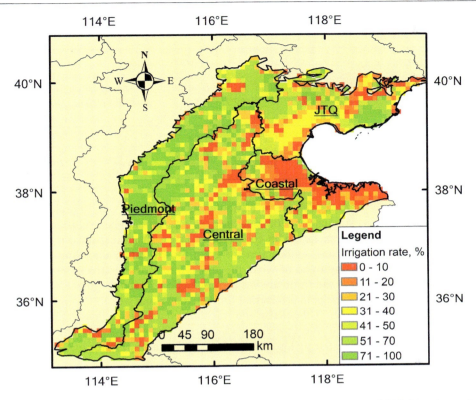

Figure 1.3 Irrigated area as a percentage of land area in the NCP circa 2005 based on global map of irrigation areas (Siebert et al., 2013).

decline began around 1957 and lasted for about five years until 1962. The second period of continuous decline began in 1980 and lasted for about five years. This decline in grain crops planting area could be explained by the fact that the farmers tended to grow more economic crops than grain after the reform of land management system, that is, replacing the collective operation system of the people's commune with the household contract responsibility system that started in 1978. The third period of decline began in 1999 and lasted for five years until

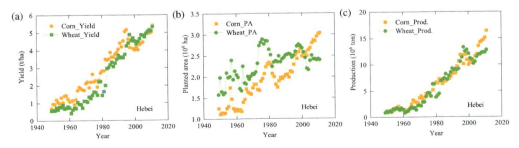

Figure 1.4 Temporal variations for maize and wheat yields (a), planted areas (b) and production (c) during the past six decades (Hebei Province, 1949–2012).

Table 1.1 Maize and wheat production during the past six decades (Hebei Province, 1949–2012)

Periods	Maize			Wheat		
	Planted area (mha/year)	Production (mton/year)	Yield (t/ha/year)	Planted area (mha/year)	Production (mton/year)	Yield (t/ha/year)
1951–1960	1.27	1.5	1.2	1.94	1.4	0.7
2001–2010	2.74	12.7	4.6	2.39	11.5	4.8
Change	115%	759%	300%	23%	734%	577%

2004, stemmed mainly from the release of National Regulations on Grain Purchase on June 6, 1998. The Regulations stipulate that the purchases of grain should be exclusive to national grain enterprises and their branches, with conservation prices. An unexpected result was the Regulations limiting the farmers' enthusiasm for grain growth rather than stimulating it. In terms of the overall increasing trend of yield, the above three sowing area reduction periods were all reflected, especially the third sowing area reduction period (Figure 1.4). The same situation is reflected in the change in production for both maize and wheat (Figure 1.4). During the past six decades (1949–2012), the maize and wheat planted area increased 115% and 23%, respectively, but the maize and wheat production increased 759% and 734%. The main reason of this dissymmetry was that the yields increased by 300% and 577% for maize and wheat, respectively (Table 1.1).

1.3 Decrease of groundwater level in the NCP

The NCP is located in the semi-humid area of northern China, with optimum temperature for crop growth (mean annual air temperature was 13.1°C) and proper land fertility for cultivation (soil organic carbon: 5.6 ± 0.4 kg SOC/m^2, and the cultivated land shares 54% of the total land). However, the precipitation is limited (about 525 mm annual mean precipitation), and there is significant water resource gap between crop water consumption (about 710 mm annual evapotranspiration for the domestic double cropping system) and natural precipitation. The NCP accounted for 25% of wheat and 18% of maize production in China within the latest decade (2002–2011). Therefore, the NCP is also considered as the "Grain Basket" in China and as the global hotspots of groundwater depletion (Figure 1.5a), as groundwater is the primary source of irrigation water. The depth to groundwater level (DTW) continuously decreased to about 50 m in the piedmont area of the NCP in 2020 (Figure 1.5b), where the DTW was only <10 m in the 1970s. The main reason for groundwater depletion is the nature of intensive agriculture with high-level irrigation.

During the past 40 years in the NCP, Taihang Mountain piedmont region experienced the most severe groundwater depletion, especially around the areas surrounding Beijing, Baoding, Shijiazhuang and Xingtai, where the DTW reached more than 40 m. On the one hand, except for the areas in north-western Shandong Province and northern Henan Province where most of the farmland is irrigated by the Yellow River, the spatial characteristics of groundwater decline highly correlate with the spatial pattern of grain production in the NCP (Pei et al., 2015); the high wheat- and maize-producing areas in the NCP closely coincide with the areas

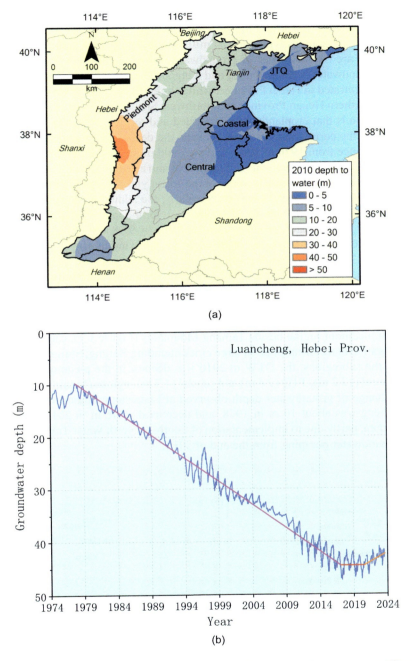

Figure 1.5 Spatial distribution in DTW in NCP in 2010 (a) and changes of DTW at Luancheng Agro-Ecosystem Experimental Station, Chinese Academy of Sciences, during 1970–2023 (b).

with the highest groundwater consumption in the middle of the piedmont plain. On the other hand, although the yield of wheat and maize around Beijing is moderate, it has the heaviest burden of groundwater resources with over 20 million population. In Beijing, domestic water and industrial water are the main uses of groundwater, which is different from the excessive consumption of groundwater in Shijiazhuang and Xingtai. The groundwater consumption of the NCP is mainly concentrated in Hebei Province, while slight decrease of groundwater appears in the plain areas of northern Henan Province and northwest Shandong Province. The main reason for that slight decline is that cropland can be irrigated and receives recharge from the Yellow River. According to the correlation observed between the decline in groundwater levels in the cities of Hebei Province from 1970 to 2008, and the yield of maize and wheat in 2011, there is a significant positive correlation with the yield of wheat and maize at the city level (Figure 1.6). The correlation shows that with 1 t/ha increase of the wheat yield, the additional decrease in the rate of groundwater level will be 0.33 m/year. Similarly, the additional decrease in the rate of groundwater level will be 0.28 m/year for 1 t/ha increase for the maize yield (Figure 1.6).

It is estimated that groundwater resources were depleted by 92.8 km³ throughout NCP (about 20 mm/year, 1980–2011, Table 1.2) based on groundwater level monitoring and modelling analyses (Cao et al., 2013). Groundwater depletion is simulated higher in the piedmont (51.2 km³/year, or 45 mm/year), lower in the central (35.2 km³/year, or 16 mm/year) and much lower in the coastal (3.2 km³/year, or 4 mm/year) and TTQ (the city group including Tianjin, Tangshan and Qinhuangdao, 3.2 km³/year, 5 mm/year) regions (Table 1.2 and Figure 1.7), similar to the spatial distribution in irrigated areas (Figure 1.2). The greatest DTW in the NCP is located in the central and northern piedmont region beside Taihang Mountains (Figure 1.7), caused by intensive pumping for irrigation and consumption in large cities including Beijing, Shijiazhuang, Baoding and Xingtai. In the subregions, the DTW in 2010 was deepest in the piedmont (23.5 m), followed by the central (13.2 m), TTQ (7.4 m) and coastal (4.7 m) regions (Figure 1.7). Figure 1.7b illustrated the change of groundwater depth observed at Luancheng Station from 1974–2023. DTW at Luancheng was about 10 m in 1978, and decreased to 46 m in 2016, and gradually recovered after then, partly due to implementation of South–to North Water Transfer Project and limitation of groundwater pumping from the end of 2014.

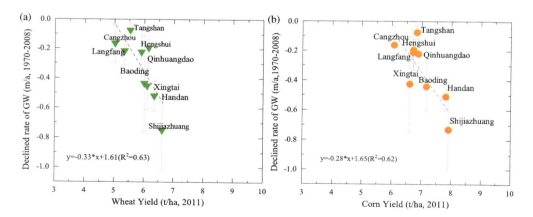

Figure 1.6 Spatial correlations between groundwater depletion and the yields of wheat (a) and corn (b) in the NCP.

Population growth, food demand and water crisis   11

Table 1.2 Groundwater depletion in the NCP during 1980–2011

Regions	Groundwater depletion (1980–2011, km³/year)	Groundwater depletion (km³/year)	Groundwater depletion (mm/year)	Specific yield	Groundwater level decline (m/year)
NCP	92.8	2.9	20	0.09	0.23
Piedmont	51.2	1.6	45	0.10	0.43
Central	35.2	1.1	16	0.08	0.21
Coastal	3.2	0.1	4	0.07	0.06
TTQ	3.2	0.1	5	0.06	0.08

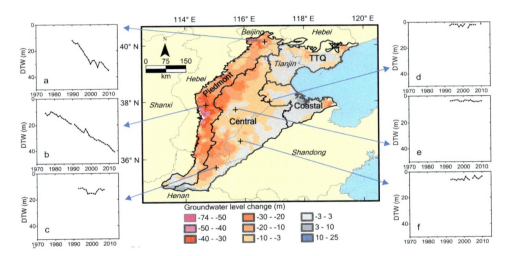

Figure 1.7 Groundwater level change with representative hydrograph showing changes in DTW during the rapid agricultural development period in the NCP (1970–2008).

The groundwater pumpage was increasing continually from the 1950s to 2000 (Figure 1.8) (Cao et al., 2013; Zhang et al., 2011), combined with the intensive agriculture development and population boom in the NCP. However, the water supply (especially the groundwater) in the NCP decreased asynchronously while the population boomed during the past 20 years, except for a slight increase in Beijing and Tianjin (Figure 1.9a and b, 0.1 billion m³/year in Beijing and 2.6 billion m³/year in Tianjin). During the past 20 years, the total water supply for the three main provinces (Beijing, Tianjin and Hebei) in the NCP decreased by 0.12 billion m³/year (2000–2020) (Figure 1.9d). The main reason for this trend was the irrigation water (mainly sourced from groundwater) decreased in Hebei Province via reduced wheat planted area. Meanwhile, the water supply structure had significant change in the past 20 years. The main change was that groundwater was replaced by water from the South-to-North Water Diversion Project (SNWDP), especially in Beijing and Tianjin. The water transferred from the SNWDP (including the eastern route and the central route) has accumulated to 50 billion m³ from

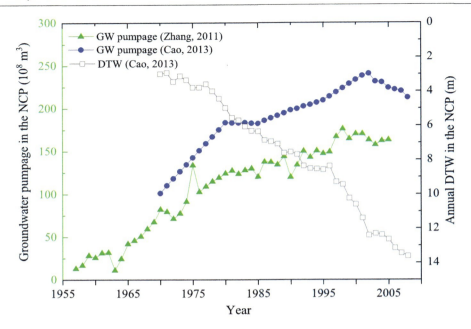

Figure 1.8 Estimated irrigation pumpage and DTW in the NCP from 1957 to 2008.

December 12, 2014, until January 7, 2022, including 5.3 billion m³ of water transferred from the eastern route and 44.7 billion m³ from the central route (Ministry of Water Resources of the People's Republic of China, 2022). The transferred water by the SNWDP has fed 140 million people with safe water, and replaced about 60% groundwater pumpage in Beijing and Tianjin (Figure 1.9d). Benefiting from the ecological water replenishment for rivers by the SNWDP, the runoff of Yongding River in Beijing (~865 km) recovered for the first time since 1965 (Ministry of Water Resources of the People's Republic of China, 2022).

1.4 Degradation of groundwater quality in the NCP

Nitrate (NO_3^-) is one of the main groundwater pollutants in agricultural regions. Agricultural land use represents the largest diffuse pollution threat to groundwater quality on a global scale (Haller et al., 2013). High nitrate concentration in groundwater leads to public health risk and environmental pollution, which has become a common problem in many parts of the world. In the NCP, limited groundwater quality data preclude detailed evaluation of agricultural impacts. The most comprehensive evaluation included 295 samples (1998–2000) which showed that 13% (38 out of 295) of wells exceeded the US EPA Maximum Contaminant Level (MCL) of 10 mg NO_3-N/L, and the highest sites located in the piedmont (26 out of 61 samples) (Chen et al., 2005). Unsaturated zone sampling coupled with crop yield and fertilizer data indicates nitrate accumulation in deep soils and high potential for nitrate leaching, particularly during the summer maize season in response to intense rains (Fang et al., 2006). Limited data (36 samples) from a survey in the lower coastal plain showed that NO_3-N in shallow groundwater averaged 24.1 mg NO_3-N/L, and 60% (22 of 36) of samples exceeded the MCL (Han et al., 2014).

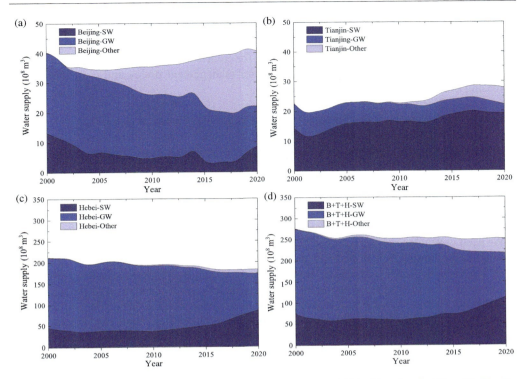

Figure 1.9 Water supply structure in the NCP during 2000–2020. (a) is for Beijing, (b) Tianjin, (c) Hebei, and (d) for the total of the three province/municipalities, B is short for Beijing, T for Tianjin, H for Hebei; SW and GW are for Surface and Groundwater, respectively. The water supply source "Other" is mostly from the SNWDP.

A synoptic survey (27 samples) in the southern region reveals that groundwater nitrate was related to field fertilization, irrigation and water level depth in a shallow groundwater region adjacent to the Yellow River (Shen et al., 2011). The latest research suggests that land use change was the major contributing factor to different nitrate concentrations beneath the groundwater. It showed that 80% of samples from residential areas, 49% of samples from farmland, 86% of samples from economic forestland and 6.9% of samples from natural vegetation exceeded the WHO standard (Wang et al., 2017).

Extensive use of irrigation leads not only to low water but also to low fertilizer use efficiency; as a result, considerable nutrients are leached out of root zone and move downward gradually, to finally contaminate groundwater. The investigation from pore sampling in different land use types implied that huge amount of nitrate is accumulated in the thickening vadose zone in NCP (Liu, 2023). The nitrate-nitrogen accumulated in an entirely 45 m unsaturated zone at a wheat-maize double crop grain field can reach up to 6586.7 kg N/ha (Liu et al., 2022). It is even much higher at the more intensive cultivated lands, such as vegetables and fruit orchards. Sustainable management of groundwater in the NCP, therefore, includes not only the quantity aspect, namely, preventing overdraft, but also the quality aspect of preventing pollution from agricultural activities.

1.5 Challenges for sustainable groundwater management

Generally, there are three critical challenges about groundwater management in the NCP: (1) dilemma in balancing water for food security and for environmental restoration, (2) innovation and practice of efficient and effective groundwater management policy and (3) coping with increasing drought risks under a warming climate.

1. Dilemma in balancing water for food security and for environmental restoration.

    Associated with groundwater table dropping regionally in the NCP, almost all the rivers run dry and become seasonal only in the rainy season. At the same time, the wetland has greatly shrunk to 1/10th of its area in the 1950s, with loss of its most ecological functions. All the severe ecological degradation comes from subsequent pumping of groundwater for irrigation to obtain higher yield. Even though some positive phenomenon appears after the SNWD project is implemented, the dilemma of using groundwater for high food yield or conserve the water resources for ecological or environmental restoration is still a big challenge today. We need to develop more efficient water use technologies on the one hand, but the segmented small-scale farming system restricts the application of new technologies in many aspects.

    To use the limited water resource more efficiently, it is urgently needed to develop integrated measures for exploring the potential of water productivity at farmland level to produce more grains per drop of water. Application of precision irrigation or even fertigation could reduce deep percolation significantly and prevent groundwater from contamination. But the fragmented small farming system causes high cost of inducing these technologies, without mentioning the lack of knowledge and skills of most farmers. Besides, better planning is also useful for improving water use efficiency and help achieve a more coordinated land, water, food system.

2. Innovation and practice of efficient and effective groundwater management policies.

    The fundamental problem at a practical level in groundwater extraction is its spontaneity and disorganization. It is challenging to manage the groundwater extraction in an orderly manner and use it efficiently as groundwater is extracted by millions of small farmers from hundreds of thousands of wells. Some new soft approaches, such as water rights, water metering, were applied and tested in the NCP. The effects on economical use of water are significant, some of them were transitory due to the complexity in practice or unconvincing metering method, such as estimating water extraction by electricity consumption. Under the background of rising trend of collective land operation in recent years, new groundwater governing policies and associated smart technologies are greatly expected to be innovated and applied in practice in NCP to achieve sustainable groundwater management.

3. Coping with increasing drought risks under a warming climate.

    Climate change markedly influences the budget of groundwater resource through altering the components of water recycling in the NCP. First, the changes in the timing, frequency and intensity of precipitation directly or indirectly influence the groundwater recharge and agricultural irrigation demand. Second, the rising CO_2 concentration and temperature change the water use efficiency of crops and potential evapotranspiration, consequently increasing the water demand of crops and irrigation amount. The increasing frequency of extreme events, such as flood, drought and heat wave, also exacerbates the uncertainties of groundwater recharge and consumption. Overall, the negative effects of climate change on crop growth and water resource far outweigh the positive effects. The groundwater resource in a warming and thirsty environment under ongoing climate change is difficult to meet the increase in food demand due to the increasing population.

References

Cao G, Zheng C, Scanlon B R, Liu J, Li W. 2013. Use of flow modeling to assess sustainability of groundwater resources in the North China Plain. *Water Resources Research*, 49: 159–175.

Chen J, Tang C, Sakura Y, Yu J, Fukushima Y. 2005. Nitrate pollution from agriculture in different hydrogeological zones of the regional groundwater flow system in the North China Plain. *Hydrogeological Journal*, 13: 481–492.

de Fraiture C, Wichelns D, Rockström J, Kemp-Benedict E, Eriyagama N, Gordon L J, Hanjra M A, Hoogeveen J, Huber-Lee A, Karlberg L. 2007. Looking ahead to 2050: scenarios of alternative investment approaches. *Eric Kemp-Benedict*, 7(5722): S19.

Falkenmark M. 2007. Shift in thinking to address the 21st century hunger gap: moving focus from blue to green water management. *Water Resources Management*, 21(1): 3–18.

Fang Q, Yu Q, Wang E, Chen Y, Zhang G, Wang J, Li L. 2006. Soil nitrate accumulation, leaching and crop nitrogen use as influenced by fertilization and irrigation in an intensive wheat-maize double cropping system in the North China Plain. *Plant and Soil*, 284(1–2): 335–350.

Foley J A, Ramankutty N, Brauman K A, Cassidy E S, Gerber J S, Johnston M, Mueller N D, O'Connell C, Ray D K, …, Zaks D P M. 2011. Solutions for a cultivated planet. *Nature*, 478(7369): 337–342.

Haller L, Mc Carthy P, O'Brien T., et al. 2013. Nitrate pollution of groundwater. Alpha Water Systems INC.

Han D, Song X, Currell M J, Yang J, Xiao G. 2014. Chemical and isotopic constraints on evolution of groundwater salinization in the coastal plain aquifer of Laizhou Bay, China. *Journal of Hydrology*, 508: 12–27.

Hanjra M A, Qureshi M E. 2010. Global water crisis and future food security in an era of climate change. *Food Policy*, 35(5): 365–377.

Liu M. 2023. *Water and nitrate transport and its controlling mechanism in the deep vadose zone of farmland, the North China Plain*. Beijing, China: University of Chinese Academy of Sciences.

Liu M, Min L, Wu L, Pei H, Shen Y. 2022. Evaluating nitrate transport and accumulation in the deep vadose zone of the intensive agricultural region, North China Plain. *Science of the Total Environment*, 825, 153894.

Min L, Shen Y, Pei H. 2015. Estimating groundwater recharge using deep vadose zone data under typical irrigated cropland in the piedmont region of the North China Plain. *Journal of Hydrology*, 527: 305–315.

Molden D. 2007. Water responses to urbanization. *Paddy and Water Environment*, 5(4): 207–209.

Molden D, Lautze J, Shah T, Bin D, Giordano M, Sanford L. 2010. Governing to grow enough food without enough water-second best solutions show the way. *International Journal of Water Resources Development*, 26(2), 249–263.

Pei H, Min L, Qi Y, Liu X, Jia Y, Shen Y, Liu C. 2017. Impacts of varied irrigation on field water budgets and crop yields in the North China Plain: rainfed vs. irrigated double cropping system. *Agricultural Water Management*, 190: 42–54.

Pei H, Scanlon B R, Shen Y, Reedy R C, Long D, Liu C. 2015. Impacts of varying agricultural intensification on crop yield and groundwater resources: comparison of the North China Plain and US High Plains. *Environmental Research Letters*, 10(4): 1–25.

Popkin B M. 2003. The nutrition transition in the developing world. *Development Policy Review*, 21(5–6): 581–597.

Postel S L, Wolf A T. 2001. Dehydrating conflict. *Foreign Policy*, 126: 60.

Rosegrant M W, Cline S A. 2003. Global food security: challenges and policies. *Science*, 302(5652): 1917–1919.

Shen Y, Chen Y. 2010. Global perspective on hydrology, water balance, and water resources management in arid basins. *Hydrological Processes*, 24(2): 129–135.

Shen Y, Lei H, Yang D, Kanae S. 2011. Effects of agricultural activities on nitrate contamination of groundwater in a Yellow River irrigated region. *Water Quality: Current Trends and Expected Climate Change Impacts*, 348: 73–80.

Shen Y, Oki T, Kanae S, Hanasaki N, Utsumi N, Kiguchi M. 2014. Projection of future world water resources under SRES scenarios: an integrated assessment. *Hydrological Sciences Journal*, 59(10): 1775–1793.

Siebert S, Burke J, Faures J M, Frenken K, Hoogeveen J, Döll P, Portmann F T. 2010. Groundwater use for irrigation a global inventory. *Hydrology and Earth System Sciences*, 14(10): 1863–1880.

Siebert S, Henrich V, Frenken K, Burke J. 2013. Update of the digital global map of irrigation areas to version 5. Rheinische Friedrich-Wilhelms-Universität, Bonn, Germany and FAO, Rome, Italy.

The State Council of the People's Republic of China. 2022. 50 billion cubic meters of water has been transferred from the first phase of the South-to-North Water Transfer Project. https://www.gov.cn/xinwen/2022-01/08/content_5667043.htm

The World Bank. 2023. Renewable internal freshwater resources per capita. https://data.worldbank.org/indicator/ER.H2O.INTR.PC?most_recent_value_desc=true.

Tian Y. 2010. Research on the agriculture water price in Yellow River irrigation area of Shandong Province. Jinan, China: Shandong University.

Tilman D, Cassman K G, Matson P A, Naylor R, Polasky S. 2002. Agricultural sustainability and intensive production practices. *Nature*, 418(6898): 671–677.

UNDP, 2007. Human development report 2006 - beyond scarcity: power, poverty and the global water crisis. New York, USA: United Nations Development Programme.

United Nations. 2022. World population prospects 2022: summary of results. https://www.un.org/development/desa/pd/content/World-Population-Prospects-2022.

Wang S, Zheng W, Currell M, Yang Y, Zhao H, Lv M. 2017. Relationship between land-use and sources and fate of nitrate in groundwater in a typical recharge area of the North China Plain. *Science of the Total Environment*, 609: 607–620.

Water Resources Bureau of Beijing, 2021. Beijing Water Resources Bulletin. Beijing.

Water Resources Bureau of Tianjin, 2021. Tianjin Water Resources Bulletin. Tianjin.

Water Resources Bureau of Hebei, 2021. Hebei Water Resources Bulletin. Shijiazhuang.

Xiao D, Liu D, Wang B, Feng P, Bai H, Tang J. 2020. Climate change impact on yields and water use of wheat and maize in the North China Plain under future climate change scenarios. *Agriculture Water Management*, 238: 106238.

Xiao D, Shen Y, Qi Y, Moiwo J P, Min L, Zhang Y, Guo Y, Pei H. 2017. Impact of alternative cropping systems on groundwater use and grain yields in the North China Plain Region. *Agricultural Systems*, 153: 109–117.

Yuan Z, Shen Y. 2013. Estimation of agricultural water consumption from meteorological and yield data: a case study of Hebei, North China. *PLoS One*, 8(3): 1–9.

Zhang G, Lian Y, Liu C, Yan M, Wang J. 2011. Situation and origin of water resources in short supply in North China Plain. *Journal of Earth Sciences and Environment*, 32(2): 172–176. (In Chinese with English Abstract).

Chapter 2

Integrated paths to sustainable groundwater management

Yanjun Shen and Dengpan Xiao

2.1 Introduction

The North China Plain (NCP) is one of the largest alluvial plains and densely populated regions in China (Wang et al., 2019), and is one of the most important grain-producing areas (Xiao et al., 2017). With only 3% of national land and 3% of total water resources, the NCP produces about 20% of China's food grains annually (Yuan and Shen, 2013). However, water scarcity, for a long time, has been a key factor limiting sustainable agricultural development there (Liu and Wei, 1989). As a consequence, groundwater pumping became a unique alternative to support the local agricultural development. According to historical records, the number of wells dug for irrigation increased from 76.6 thousand wells in 1965 to 964.5 thousand wells in 2010 in Hebei Province. Today, millions of wells are spread over the land, averaging 4.8 wells per km^2 across the NCP (Shang et al., 2016). The rapid increase in groundwater wells led to swift evolution of agricultural intensification in just a few decades (Liu et al., 2011). Since the 1970s, to meet the increased food demand, the pattern of crop cultivation in the NCP has gradually been adjusted from two/three crops in two years into two crops per year, and locally, special winter wheat-summer maize double cropping system has continued until today (Xiao et al., 2013). This high-intensity agricultural pattern has achieved huge grain output over the past decades in the NCP (Xiao and Tao, 2014; Wang et al., 2012), surely, which attributed partly to the region's suitable climate and soil conditions (Yang et al., 2010).

Due to the competent irrigated condition, ~85% of total cropland can be irrigated across the NCP (Pei et al., 2015a), which played a crucial role in improving crop productivity even during precipitation deficit period (Liu et al., 2001). Full irrigation has proved to decouple the relationship between crop yield and precipitation variability in most parts of the NCP (Sun et al., 2006). In addition, improving fertilizer management and application is another key practice for achieving high crop yields (Fang et al., 2006). For fertilizer management and application, nitrogen (N) is the main fertilizer (59% of total fertilizer application) in the NCP, followed by P as P_2O_5 (27%) and K as K_2O (14%) (Dai et al., 2013). Generally, N applied to wheat is about 1.5 times greater than that applied to maize, and as a result, redundant N from wheat stored in soils, in such double cropping systems, could be subsequently utilized by maize (Pei et al., 2015b).

Although the double cropping system has enhanced grain production, it was at the expense of severe groundwater depletion in the plain (Fang et al., 2010). Since precipitation is difficult to form streamflow and other types of surface water which can be used for irrigation in such a region of flat terrain over the NCP, the groundwater becomes the most important source of irrigation water (Hu et al., 2010). Over the past decades, the agricultural intensification practices combined with the mismatch between huge food requirements and scarce water resources'

DOI: 10.1201/9781003221005-3

availability in NCP have led to a sharp decline in groundwater table, rivers drying up, and deterioration of water quality (Yang and Tian, 2009).

In fact, irrigated water is mainly used to meet the demand of winter wheat, as water demand (450 mm/year) of winter wheat far exceeds the precipitation (130 mm/year, as the probability of 50%) during its growth period (Sun et al., 2015). It is estimated that irrigated water must reach about 200–450 mm/year to maintain general crop yields under the double cropping system (Xiao et al., 2017). Consequently, use of irrigated water for crops is the main factor causing groundwater depletion in the NCP (Zhang et al., 2011). In general, irrigation accounts for over 70% of total groundwater use in the floodplains and 87.2% in the piedmont plain (Yang et al., 2010). Due to over-exploitation of groundwater, the groundwater depth has declined continuously from around 10 m in the beginning of the 1970s to 46 m in 2015 in the piedmont plain, where a giant deep vadose zone has grown (Pei et al., 2017). Those pose a challenge to sustainable utilization of regional groundwater resource (Moiwo et al., 2010). Notably, the rapid decline in groundwater worsened the local environmental conditions such as land subsidence and groundwater depression (Guo et al., 2015; Shang et al., 2016), which in turn threatened sustainable agricultural production in the NCP (Zhong et al., 2017).

Long-term N fertilizer use has led to pronounced agricultural success in the past few decades in the NCP (Zhang et al., 2018). The rate of N application was estimated to reach 350–600 kg/ha to maintain a high yield (Pei et al., 2015a). As a result, the average annual nitrate-N leaching loss reached 15.8 kg/ha/year for the conventional farming fertilization treatment in the NCP (Yang et al., 2015). The amount of N input determines N concentration degree in the groundwater and cropland soils, which is an important factor threatening water security (Yang et al., 2015; Zhang et al., 2018). Overall, the agriculture in the NCP is under an unsustainable development, including both long-term irrigation-induced groundwater depletion and nitrate pollution due to fertilizer use.

2.2 Contradiction between food security and water security

It is projected that global population will increase from 6.9 billion in 2010 to 10.0 billion by 2060 (Tilman et al., 2011). Food security, now and in the future, is always one of the most concerning issues globally. China occupies 20% (1.4 billion) of the global population but owns only 8% of global arable land; moreover China's agriculture is often threatened by extreme climate (Piao et al., 2010; Huang et al., 2015). Prior to the 1970s, both the Yangtze River Delta and Pearl River Delta in South China were the major grain-producing areas of China due to their abundant water resources, and have contributed to the high proportion of grain export into water-deficit North China. Owing to rapid industrialization and urbanization from the 1980s, the agricultural importance of South China decreased a lot and the NCP has become the most important granary and food basket in China in the past four decades. Groundwater depletion and its subsequent ecological degradations in the NCP stem from the mismatch between its significant role in the national food security and the relatively limited availability of water resources, which is insufficient to sustain the region's food production mission.

As mentioned previously, the NCP is currently facing a serious crisis of groundwater over-exploitation and pollution as a result of prolonged accumulation of negative effects from agricultural intensification (Hu et al., 2005). The issue has drawn global scholars' attention (Huang et al., 2015). In order to fill the gap of irrigation water usage in the NCP, the Chinese government began planning to construct the ambitious South-to-North Water Diversion Project (SNWDP) since the 1980s. And the project of middle route had been completed and implemented since

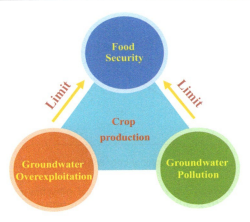

Figure 2.1 The relationship between food security and groundwater over-exploitation and pollution.

December 2014. Nowadays, the concern is how to coordinate the trade-off between food security and groundwater over-exploitation, and pollution, in the NCP (Figure 2.1) both academically and politically (Gao et al., 2015; Zhong et al., 2017).

2.2.1 Challenges of food security

Related study has noted that in the NCP large amounts of virtual water, especially blue water (i.e., extracted groundwater), are exported through cross-regional food trade, away from the NCP (Ren et al., 2018). This posed a further challenge for water shortage. How to sustain agricultural development by minimizing the negative effects of water use is the most fundamental question here (Ren et al., 2018). Finding a way of both ensuring enough food output and reducing agricultural water use is key to solve the challenge of food security (Zhang et al., 2018). Generally, the effective way to reduce blue water loss and alleviate groundwater depletion is to reduce the planting of water-consuming crops (i.e., winter wheat, vegetables, fruit trees, etc.) (Xiao et al., 2017). As a result, it may bring certain challenges and risks to regional and even national food security.

2.2.2 Challenges of groundwater conservation

The long-term groundwater over-exploitation has made the NCP one of the world's largest groundwater depletion zones, and resulted in incalculable damage to the environment (Liu et al., 2001; Moiwo et al., 2010; Shi et al., 2011; Guo et al., 2015). Especially in the piedmont plain area, the groundwater depth has severely decreased, mainly due to excessive groundwater pumping for wheat irrigation in dry seasons (Chen et al., 2003). A related study indicated that groundwater was reducing at a rate of 17.8 mm/year in the NCP during the years 1971–2015, by combining the Gravity Recovery and Climate Experiment (GRACE) with in situ observations (Gong et al., 2018). This situation will be worse in the drought years (Sun et al., 2019).

In recent decades, with the increase of regional irrigation intensity, through pumping groundwater, to ensure high and stable grain yield in the NCP, the production benefits (PB) has

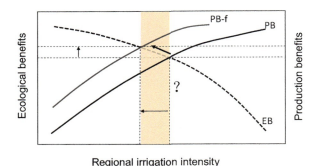

Figure 2.2 Relationship between curves of Ecological benefits (EB) and production benefits (PB) in response to groundwater irrigation intensity in the North China Plain. PB-f refers to the curve of production benefits in future with improvement of water use efficiency.

increased significantly, while the ecological benefits (EB) has declined sharply (Figure 2.2), leading to a series of ecological problems mainly stemmed from groundwater depletion. To some extent, choosing the appropriate irrigation intensity is the result of balancing PB and EB. Therefore, we should not only reduce the irrigation intensity, but also need to improve the PB through effective measures, such as improving the irrigation efficiency or increasing the water use efficiency (WUE) of crops. Overall, the main goal for the future is not only to increase EB by reducing irrigation intensity without reducing the present production benefits (PB), but also to improve it to a level of future production benefits (PB-f) through a series of management measures (Figure 2.2). This goal will help to achieve a better future with respect to harmonizing agriculture and ecosystems.

Early studies have emphasized that WUE is an important aspect in water-saving agriculture and practical measures, which includes various water-saving technologies such as deficit irrigation, critical period irrigation, straw mulch or biodiversity improvement (Li et al., 2010; Pei et al., 2015a; Fang et al., 2018). It should be noted that improving only the irrigated WUE on farmland is hard to achieve the balance of regional water use (Yan et al., 2015). Sun et al. (2015) found that under current cropping systems, even a minimum of one irrigation event per crop life cycle can cause the water table to decline. Overall, the current winter wheat-summer maize double cropping system is not sustainable in the NCP because precipitation is not sufficient for groundwater recharge under any irrigation system (Wang et al., 2008). Thus, reducing irrigated areas or cultivated areas has been regarded as an effective way of resolving water shortages (Gao et al., 2015; Meng et al., 2012; Sun et al., 2019; Xiao et al., 2017; Yan et al., 2020; Zheng et al., 2010).

In the NCP, massive synthetic fertilizer application has significantly contributed to grain production (Min et al., 2018; Pei et al., 2015a), while fertilizer is also the primary source of high nitrate concentration in groundwater and vadose zone (Chen et al., 2005; Zhao et al., 2019). Thus, a contradiction is created between food security and environmental targets. The field data indicate a high potential for nitrate accumulation and nitrate leaching in deep soils, especially for summer maize after heavy rainfall (Fang et al., 2006). Overuse of N fertilizer and declining groundwater table have created a huge nitrate reservoir in the vadose zone in the NCP (Gai et al., 2019). Groundwater sustainability could be enhanced by reducing fertilizer application and ensuring a balanced nutrient supply, while managing nutrient demand spatially and temporally (Liu et al., 2022; Shen et al., 2011). Fertilizing amount and timing need to be optimized to match crop growth requirement and minimize runoff or leakage, yet there are currently technical barriers that need to be overcome within the NCP's agriculture sector (Li et al., 2016).

If limitations on groundwater over-exploitation lasted, the groundwater will recover rapidly. This will accelerate the risk of N pollution with increased groundwater level, which carried massive N that was previously stored in the vadose zone in the groundwater (Min et al., 2022). In the piedmont plain, there is a condition that shallow groundwater recovery made the nitrogen from the unsaturated zone enter into the aquifer, where large quantities of nitrogen are stored in the thick unsaturated zone (Min et al., 2022; Wang et al., 2019).

2.3 Integrated paths to sustainable groundwater use and agricultural development in the NCP

Over the NCP, groundwater is universally over-exploited (Gong et al., 2018). Generally, unlike shallow groundwater, deep groundwater is hard to recharge (Lu et al., 2011; Min et al., 2015). In dry seasons, the pumping water for crop production is almost from deep groundwater in the eastern low plain area (Cao et al., 2013; Wang et al., 2021). In order to sustain agricultural development in the NCP, we should consider to reduce the groundwater pumping and use the irrigated water more efficiently based on the following three points:

The first and vital one is to develop an integrated field water-saving technological system to largely improve WUE, or produce more crops with per drop of water. This is at the field level.

The second is to manage agricultural production intensity to a proper and suitable extent to minimize the excessive production of agricultural goods than the objectives to meet certain extent of demand. This includes finding a suitable cultivation intensity at the field level, ensuring precipitation and extractable groundwater can adequately meet water consumption needs, as well as regulating the regional cropping pattern to achieve a scale and structure of different crops to match with the sustainable groundwater extraction threshold at the regional level.

The third is to develop and establish effective and efficient groundwater management system, including advanced groundwater governance policy and supporting technologies developed with the emerging information techniques, such as IoT, drone, remote sensing and automatic controlling unit.

These approaches are not mutually exclusive, they could augment water-saving effects if used in combination at different levels/scales (Figure 2.3).

2.3.1 Developing integrated field water-saving technological system

Water-saving measures are critical to mitigate groundwater table drawdown and to support sustainable use. Theoretically, agricultural water saving refers to making full use of natural precipitation and irrigation water, while minimizing the loss of water in the process of water transmission, allocation, and irrigation in the farm. Thus, it is necessary to carry out water-saving irrigation and integrated water-saving measures to reduce non-productive loss. In all water use sectors, agricultural water saving has the highest potential, which is related to new agricultural water-saving technologies, integration of water and fertilizer, pipe irrigation, micro-irrigation, and spray irrigation (Zhang et al., 2016).

For crop production in the NCP, farmers usually irrigate three to five times via basin irrigation during the winter wheat growing season. To stabilize the groundwater table, new irrigation

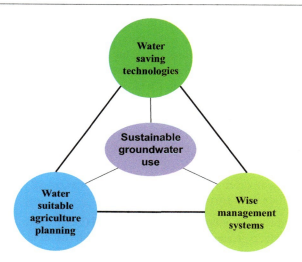

Figure 2.3 Main paths to reaching sustainable utilization of groundwater in the NCP.

strategies must be developed by increasing WUE (Sun et al., 2015). Generally, reducing irrigation amount and frequency would be a solution to alleviate and control the groundwater depletion (Pei et al., 2017; Fang et al., 2018). Moreover, different irrigation strategies, such as critical period irrigation, deficit irrigation, and minimum irrigation, could be used to improve water budgets and crop yields (Zhang et al., 2016).

Deficit irrigation is defined as the irrigated water below crop water requirements. Moderate water deficit during grain filling stage could increase mobilization of assimilates stored in vegetative tissues and grain, and thus improve grain yield and WUE (Zhang et al., 2013). It can promote crop growth and is more efficient in soil water utilization during the reproductive stage (Fang et al., 2018). Reduce irrigation times to twice that for wheat, the first at jointing and the second at heading or anthesis, which will promote higher WUE and yield (Li et al., 2010). Except for deficit irrigation measures, Fang et al. (2018) reported that increasing irrigation frequency could maintain higher water content in the topsoil layer, and improve crop yield and water use under limited water supply. A minimum irrigation strategy showed the potential to maintain the groundwater balance with higher grain production and WUE (Yan et al. 2020). Moreover, the WUE could increase by 10%–20% under minimum irrigation (irrigate only the upper root zone to field holding capacity before planting) compared to conventional irrigation practice (Zhang et al., 2006). In addition, supplemental irrigation at early growth stages of crops enables vegetation root to grow into deeper soil layer, effectively enhancing uptake of water from the deeper subsoil layers (Xu et al., 2016).

On the other side, some efficient water-saving irrigation methods, including spray irrigation, micro-irrigation, and high-standard low-pressure pipeline irrigation, are being vigorously advocated in areas with over-exploited groundwater (Zhang et al., 2016). In addition, water and fertilizer can be specifically adjusted based on crop characteristics.

To reduce irrigation without compromising crop yield, a structural measure involving planting water-saving and drought-resistant varieties with fine moisture-holding ability of soils on wheat planting land has been put into effect (Shen et al., 2013). Protective tillage is more suitable for high water consumptive crops (Pei et al., 2015a). Wheat planting area should be tilled as infrequently as possible, so as to ensure the lowest water loss by evaporation (Zhang et al., 2016). After harvesting, crop stalks can be further disposed, and if being evenly covered over harvested land surface, water-saving effect and crop yields will be enhanced (Chen et al., 2007).

The NCP is a hotspot of high crop yields and is also a region with high N and phosphorus (P) leaching (Hu et al., 2005; Zhao et al., 2019). The key to reducing agricultural N/P input into soils is to control water pollution (Liu et al., 2022). It is necessary to improve nutrient recycling rates and reduce the rate of synthetic N application to a lower level (Min et al., 2018). Due to several smallholders' community in the NCP, it is also important to equip smallholders with advanced agricultural technologies and to make them achieve higher output outside of research-oriented experiments. For instance, a comprehensive decision-support integrated soil-crop system management program that can achieve greater nitrogen use efficiency, net income, and environmental performance has been implemented (Fang et al., 2015). By a pressure device, mixed water and fertilizer can be applied via droppers and pipes using drip irrigation and micro-jet methods (Shang et al., 2016).

2.3.2 Planning proper intensity and water-suitable agricultural patterns

In the NCP, excessive food production and virtual water export placed an additional burden on local water supplies and the environment as well (Zhong et al., 2017). Thus, reducing irrigated areas and water-consuming crops, especially winter wheat, is an effective strategy to resolve water shortage (Figure 2.4). Many studies indicated that the current intensity of winter wheat-summer maize double cropping system in the NCP has far exceeded the land and water supporting capacity (Luo et al., 2018; Sun et al., 2011; Xiao et al., 2021; Zhao et al., 2018). The adjustment of planting intensity should be urgently optimized. Recently, an emerging initiative suggests taking full advantage of the yield-enhancing potential of autumn crops, and to

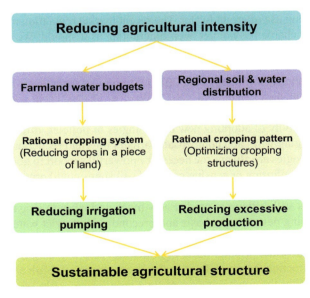

Figure 2.4 Sustainable development path of regional groundwater by reducing agricultural intensity.

minimize the water stress generated by large-scale winter wheat planting (Xiao et al., 2017). Specifically, the acreage of groundwater-irrigated winter wheat should be moderately reduced in places where deep groundwater is not available. It is better to plant crops harvested once a year (such as maize, cotton, peanuts, and sunflower) rather than winter wheat and summer maize double harvest in one year, so that one quarter of the land is left fallow and another one quarter is used for water retention. Moreover, non-agricultural crops should be introduced to replace agricultural crops where groundwater is over-exploited. For example, wheat areas can be sparsely afforested with drought-resistant and hardy as well as economically beneficial tree species. In riverside and lakeside areas, agricultural land can be restored to wetlands by planting aquatic plants (Shang et al., 2016).

In addition, studies have evaluated the effects of wheat-maize double cropping system on crop output and water consumption by field experiments and crop model simulations (Meng et al., 2012; van Oort et al., 2016). Xiao et al. (2017) noted that three crops every two years (winter wheat and summer maize followed by fallow and early maize during two consecutive years) was an effective cropping system to maintain a balance between crop yield and water consumption. Cui et al. (2019) indicated mono-cropped maize cropping system was an effective measure to alleviate current water shortage pressure but it might compromise crop yield. In other aspects, covering the planting land with straw (straw mulching) during crop grow period and/or fallow period can effectively reduce soil evaporation and conserve soil moisture (Chen et al., 2007). It was estimated that the practice of straw mulching could reduce soil evaporation by about 40% for winter wheat and by about 56% for summer maize (Zhang et al., 2016).

2.3.3 Implementing wise groundwater management systems

In the NCP, the pumping water for irrigation is currently fee-free, but the bill of electricity consumed should be paid by farmers themselves (Meng et al., 2012). This has led to farmers' unrestrained extraction of groundwater (Pei et al., 2017). Hence, water must be priced to exert a limiting role in saving use of groundwater. The recent market regulation mechanism needs to be improved, as water charge has not yet been functioning as it should. We believe that economic means could play a key role in helping solve the problem.

Currently, it is the lack of law enforcement that contributed to serious extensive use of groundwater resources in the NCP (Shang et al., 2016). Strong and strict water legislation needs to be developed and established (Zhao et al., 2015; Wang et al., 2021). In addition, socioeconomic factors are useful for extension of new technologies and policies. For example, in order to increase farmers' positivity of adopting new technologies or policies, different types of subsidies are always encouraging ways and compensate farmers for yield reduction if any.

Effective technologies are always the necessary powerful support to good policy implementation. For instance, monitoring or sensing crop and soil status could provide instant and accurate information for irrigation decision; without good knowledge and skills in crop water management modeling, it is still difficult to predict how much and when to irrigate the crops. Accurate and low-cost metering device is the basis and precondition of strict water governance policies being implemented. It is essential to obtain accurate information of water quantity for practicing the soft paths such as water pricing, water right trading, pumping quota, or even redline control of groundwater extraction. Meanwhile, it is also necessary to establish a groundwater monitoring platform including the functions of real-time monitoring; early warning, prevention, and control through sensors; remote sensing; as well as an inter-sectoral groundwater management and coordination mechanism.

2.4 Summary

To better optimize synergies and trade-offs between food security and water security in the NCP, several measures should be implemented, such as integration of water and N-coupled management, controlling groundwater quality and quantity, and transitioning from a food-oriented to environment-oriented policy. Water security and food security currently are the most important issues in the NCP. To ensure food security, strategies for improving agricultural WUE and groundwater management should be explored. The advanced water-saving technologies, agricultural best practices, and whatever protects groundwater and mitigates nitrate pollution should be on the agenda. Future crop management should consider more comprehensive factors that increase crop yields while minimizing negative impacts on water resources.

Along with further limitations on groundwater exploitation in the NCP, pollution control of groundwater will become more difficult, time-consuming, and costly compared to surface water, given that deep groundwater is currently a reserve source of safe drinking water for the residents too. In order to fundamentally prevent over-exploitation of groundwater, it is imperative to address integrated conservation of groundwater effectively from both aspects of quantity and quality.

References

Cao G, Zheng C, Scanlon B R, Liu J, Li W. 2013. Use of flow modeling to assess sustainability of groundwater resources in the North China Plain. *Water Resources Research*, 49: 159–175.

Chen J, Tang C, Sakura Y, Yu J, Fukushima Y. 2005. Nitrate pollution from agriculture in different hydrogeological zones of the regional groundwater flowsystem in the North China Plain. *Hydrogeology Journal*, 13: 481–492.

Chen J, Tang, C, Shen Y, Sakura Y, Kondoh A., Shinada J. 2003. Use of water balance calculation and tritium to examine the dropdown of groundwater table in the piedmont of the North China Plain (NCP). *Environmental Geology*, 44(5): 564–571.

Chen S, Zhang X, Pei D, Sun H, Chen S. 2007. Effects of straw mulching on soil temperature, evaporation and yield of winter wheat: field experiments on the North China Plain. *Annals of Applied Biology*, 150(3): 261–268.

Cui J, Sui P, Wright D, Wang D, Sun B, Ran M, Shen Y, Li C, Chen Y. 2019. Carbon emission of maize-based cropping systems in the North China Plain. *Journal of Cleaner Production*, 213: 300–308.

Dai X, Ouyang Z, Li Y, Wang H. 2013. Variation in yield gap induced by nitrogen, phosphorus and potassium fertilizer in North China Plain. *PLoS One*, 8(12): e82147.

Fang J, Zhou A, Ma C, Li C, Cai H, Gan Y, Liu Y. 2015. Evaluation of nitrate source in groundwater of southern part of North China Plain based on multi-isotope. *Journal of Central South University*, 22: 610–618.

Fang Q, Ma L, Green T R, Yu Q, Wang T, Ahuja L R. 2010. Water resources and water use efficiency in the North China Plain: current status and agronomic management options. *Agricultural Water Management*, 97(8): 1102–1116.

Fang Q, Yu Q, Wang E, Chen Y, Zhang G, Wang J, Li L. 2006. Soil nitrate accumulation, leaching and crop nitrogen use as influenced by fertilization and irrigation in an intensive wheat-maize double cropping system in the North China Plain. *Plant Soil*, 284: 335–350.

Fang Q, Zhang X, Shao L, Chen S, Sun H. 2018. Assessing the performance of different irrigation systems on winter wheat under limited water supply. *Agricultural Water Management*, 196: 133–143.

Gai X, Liu H, Liu J, Zhai L, Wang H, Yang B, Ren T, Wu S, Lei Q. 2019. Contrasting impacts of long-term application of manure and crop straw on residual nitrate-n along the soil profile in the North China Plain. *Science of the Total Environment*, 650(Part 2): 2251–2259.

Gao B, Ju X, Meng Q, Cui Z, Christie P, Chen X, Zhang F. 2015. The impact of alternative cropping systems on global warming potential, grain yield and groundwater use. *Agriculture, Ecosystems and Environment*, 203: 46–54.

Gong H, Pan Y, Zheng L, Li X, Zhu L, Zhang C, Zhi H, Li Z, Wang H, Zhou C. 2018. Long-term groundwater storage changes and land subsidence development in the North China Plain (1971–2015). *Hydrogeology Journal*, 26(5): 1417–1427.

Guo H, Zhang Z, Cheng G, Li W, Li T, Jiao J. 2015. Groundwater-derived land subsidence in the North China. *Environmental Earth Sciences*, 74(2): 1415–1427.

Hu K, Huang Y, Li H, Li B, Chen D, White R E. 2005. Spatial variability of shallow groundwater level, electrical conductivity and nitrate concentration, and risk assessment of nitrate contamination in North China Plain. *Environment International*, 31(6): 896–903.

Hu Y, Moiwo J P, Yang Y, Han S, Yang Y. 2010. Agricultural water-saving and sustainable groundwater management in Shijiazhuang Irrigation District, North China Plain. *Journal of Hydrology*, 393(3–4): 219–232.

Huang F, Liu Z, Ridoutt B G, Huang J, Li B. 2015. China's water for food under growing water scarcity. *Food Security*, 7(5): 933–949.

Li Q, Dong B, Qiao Y, Liu M, Zhang J. 2010. Root growth, available soil water, and water-use efficiency of winter wheat under different irrigation regimes applied at different growth stages in North China. *Agricultural Water Management*, 97(10): 1676–1682.

Li Y, Liu H, Huang G, Zhang R, Yang H. 2016. Nitrate nitrogen accumulation and leaching pattern at a winter wheat: summer maize cropping field in the North China Plain. *Environmental Earth Science*, 75(2): 118.

Liu C, Wei Z. 1989. *Agro-hydrology and water resources in the North China Plain*. Science Press: Beijing, China, pp. 61–128 (in Chinese).

Liu C, Yu J, Kendy E. 2001. Groundwater exploitation and its impact on the environment in the North China Plain. *Water International*, 26(2): 265–272.

Liu J, Cao G, Zheng C. 2011. Sustainability of groundwater resources in the North China Plain. In: Jones, J. A. A. (Ed.), *Sustaining groundwater resources, international year of planet earth*, Springer, Dordrecht., pp. 69–87.

Liu M, Min L, Wu L, Pei H, Shen Y. 2022. Evaluating nitrate transport and accumulation in the deep vadose zone of the intensive agricultural region, North China Plain. *Science of the Total Environment*, 825: 153894.

Lu X, Jin M, van Genuchten, M T, Wang B. 2011. Groundwater recharge at five representative sites in the Hebei plain, China. *Groundwater*, 49(2): 286–294.

Luo J, Shen Y, Qi Y, Zhang Y, Xiao D. 2018. Evaluating water conservation effects due to cropping system optimization on the Beijing-Tianjin-Hebei plain, China. *Agricultural Systems*, 159: 32–41.

Meng Q, Sun Q, Chen X, Cui Z, Yue S, Zhang F, Romheld V. 2012. Alternative cropping systems for sustainable water and nitrogen use in the North China Plain. *Agriculture, Ecosystems and Environment*, 146(2012): 93–102.

Min L, Liu M, Wu L, Shen Y. 2022. Groundwater storage recovery raises the risk of nitrate pollution. *Environmental Science & Technology*, 56(1): 8–9.

Min L, Shen Y, Pei H. 2015. Estimating groundwater recharge using deep vadose zone data under typical irrigated cropland in the piedmont region of the North China Plain. *Journal of Hydrology*, 527: 305–315.

Min L, Shen Y, Pei H, Wang P. 2018. Water movement and solute transport in deep vadose zone under four irrigated agricultural land-use types in the North China Plain. *Journal of Hydrology*, 559: 510–522.

Moiwo J P, Yang Y, Li H, Han S, Yang Y. 2010. Impact of water resource exploitation on the hydrology and water shortage in Baiyangdian Lake. *Hydrological Processes*, 24(21): 3026–3039.

Pei H, Min L, Qi Y, Liu X, Jia Y, Shen Y, Liu C. 2017. Impacts of varied irrigation on field water budgets and crop yields in the North China Plain: rainfed vs. irrigated double cropping system. *Agricultural Water Management*, 190: 42–54.

Pei H, Scanlon B R, Shen Y, Reedy R, Long D, Liu C. 2015a. Impacts of varying agricultural intensification on crop yield and groundwater resources: comparison of the North China Plain and US High Plains. *Environmental Research Letters*, 10(4): 044013.

Pei H, Shen Y, Liu C. 2015b. Review on nitrogen and water balance of the typical wheat and corn rotation system in the North China plain transfer. *Chinese Journal of Applied Ecology*, 26(1): 283–296 (in Chinese).

Piao S, Ciais P, Huang Y, Shen Z, Peng S, Li J, Zhou L, Liu H, Ma Y, …, Fang J. 2010. The impacts of climate change on water resources and agriculture in China. *Nature*, 467(7311): 43–51.

Ren D, Yang Y, Yang Y, Richards K, Zhou X. 2018. Land-water-food nexus and indications of crop adjustment for water shortage solution. *Science of the Total Environment*, 626: 11–21.

Shang Y, You B, Shang L. 2016. China's environmental strategy towards reducing deep groundwater exploitation. *Environmental Earth Sciences*, 75(22): 1439.

Shen Y, Lei H, Yang D, Kanae S. 2011. Effects of agricultural activities on nitrate contamination of groundwater in a yellow river irrigated region water quality. *Current Trends and Expected Climate Change Impacts*, 348, 73–80.

Shen Y, Zhang Y, Scanlon B R, Lei H, Yang D, Yang F. 2013. Energy/water budgets and productivity of the typical croplands irrigated with groundwater and surface water in the North China Plain. *Agricultural and Forest Meteorology*, 181: 133–142.

Shi J, Wang Z, Zhang Z, Fei Y, Li Y, Zhang F, Chen J, Qiao Y. 2011. Assessment of deep groundwater over-exploitation in the North China Plain. *Earth Science Frontiers*, 2(4): 593–598 (in Chinese).

Sun H, Liu C M, Zhang X, Shen Y, Zhang Y. 2006. Effects of irrigation on water balance, yield and WUE of winter wheat in the North China Plain. *Agricultural Water Management*, 85(1–2): 211–218.

Sun H, Zhang X, Liu X, Liu X, Shao L, Chen S, Wang J, Dong X. 2019. Impact of different cropping systems and irrigation schedules on evapotranspiration, grain yield and groundwater level in the North China Plain. *Agricultural Water Management*, 211(C): 202–209.

Sun H, Zhang X, Wang E, Chen S, Shao L. 2015. Quantifying the impacts of irrigation on groundwater reserve and crop production: a case study in the North China Plain. *European Journal of Agronomy*, 70: 48–56.

Sun Q, Kröbel R, Müller T, Römheld V, Cui Z, Zhang F, Chen X. 2011. Optimization of yield and water-use of different cropping systems for sustainable groundwater use in North China Plain. *Agricultural Water Management*, 98(5): 808–814.

Tilman D, Balzer C, Hill J, Befort B L. 2011. Global food demand and the sustainable intensification of agriculture. *Proceedings of the National Academy of Sciences of the United States of America*, 108(50): 20260–20264.

van Oort P A J, Wang G, Vos J, Meinke H, Li B, Huang J, van der Werf W. 2016. Towards groundwater neutral cropping systems in the alluvial fans of the North China Plain. *Agricultural Water Management*, 165: 131–140.

Wang E, Yu Q, Wu D R, Xia J. 2008. Climate, agricultural production and hydrological balance in the North China Plain. *International Journal of Climatology*, 28: 1957–1970.

Wang J, Wang E, Yang X, Zhang F, Yin H. 2012. Increased yield potential of wheat-maize cropping system in the North China Plain by climate change adaptation. *Climatic Change*, 113(3): 825–840.

Wang K, Chen H, Fu S, Li F, Wu Z, Xu D. 2021. Analysis of exploitation control in typical groundwater over-exploited area in North China Plain. *Hydrological Sciences Journal*, 66(5): 851–861.

Wang S, Hu Y, Yuan R, Feng W, Pan Y, Yang Y. 2019. Ensuring water security, food security, and clean water in the North China Plain-conflicting strategies. *Current Opinion in Environmental Sustainability*, 40: 63–71.

Xiao D, Liu D, Feng P, Wang B, Waters C, Shen Y, Qi Y, Bai H, Tang J. 2021. Future climate change impacts on grain yield and groundwater use under different cropping systems in the North China Plain. *Agricultural Water Management*, 246: 106685.

Xiao D, Tao F. 2014. Contributions of cultivars, management and climate change to winter wheat yield in the North China Plain in the past three decades. *European Journal of Agronomy*, 52: 112–122.

Xiao D, Shen Y, Qi Y, Moiwo J P, Min L, Zhang Y, Guo Y, Pei H. 2017. Impact of alternative cropping systems on groundwater use and grain yields in the North China Plain Region. *Agricultural Systems*, 153(C): 109–117.

Xiao D, Tao F, Liu Y, Shi W, Wang M, Liu F, Zhang S, Zhu Z. 2013. Observed changes in winter wheat phenology in the North China Plain for 1981–2009. *International Journal of Biometeorology*, 57(2): 275–285.

Xu C, Tao H, Tian B, Gao Y, Ren J, Wang P. 2016. Limited-irrigation improves water use efficiency and soil reservoir capacity through regulating root and canopy growth of winter wheat. *Field Crops Research*, 196: 268–275.

Yan N, Wu B, Perry C, Zeng H. 2015. Assessing potential water savings in agriculture on the Hai Basin plain, China. *Agricultural Water Management*, 154: 11–19.

Yan Z, Zhang X, Rashid M A, Li H, Jing H, Hochman Z. 2020. Assessment of the sustainability of different cropping systems under three irrigation strategies in the North China Plain under climate change. *Agricultural Systems*, 178: 102745.

Yang X, Lu Y, Tong Y, Yin X. 2015. A 5-year lysimeter monitoring of nitrate leaching from wheat-maize rotation system: comparison between optimum N fertilization and conventional farmer N fertilization. *Agriculture, Ecosystems & Environment*, 199: 34–42.

Yang Y, Tian F. 2009. Abrupt change of runoff and its major driving factors in Haihe River Catchment, China. *Journal of Hydrology*, 374(3): 373–383.

Yang Y M, Yang Y H, Moiwo J P, Hu Y. 2010. Estimation of irrigation requirement for sustainable water resources reallocation in North China. *Agricultural Water Management*, 97(11): 1711–1721.

Yuan Z, Shen Y. 2013. Estimation of agricultural water consumption from meteorological and yield data: a case study of Hebei, North China. *PLoS One*, 8(3): e58685.

Zhang Q, Wang H, Wang L. 2018. Tracing nitrate pollution sources and transformations in the over-exploited groundwater region of North China using stable isotopes. *Journal of Contaminant Hydrology*, 218: 1–9.

Zhang X, Chen S, Sun H, Shao L, Wang Y. 2011. Changes in evapotranspiration over irrigated winter wheat and maize in North China Plain over three decades. *Agricultural Water Management*, 98(6): 1097–1104.

Zhang X, Pei D, Chen S, Sun H, Yang Y. 2006. Performance of double cropped winter wheat-summer maize under minimum irrigation in the North China Plain. *Agronomy Journal*, 98(6): 1620–1626.

Zhang X, Qin W, Xie J. 2016. Improving water use efficiency in grain production of winter wheat and summer maize in the North China Plain: a review. *Frontiers of Agricultural Science and Engineering*, 3(1): 25–33.

Zhang X, Wang Y, Sun H, Chen S, Shao L. 2013. Optimizing the yield of winter wheat by regulating water consumption during vegetative and reproductive stages under limited water supply. *Irrigation Science*, 31(5): 1103–1112.

Zhao X, Liu J, Liu Q, Tillotson M R, Guan D, Hubacek K. 2015. Physical and virtual water transfers for regional water stress alleviation in China. *Proceedings of the National Academy of Sciences of the United States of America*, 112(4): 1031–1035.

Zhao Z, Qin W, Bai Z, Ma L. 2019. Agricultural nitrogen and phosphorus emissions to water and their mitigation options in the Haihe Basin, China. *Agricultural Water Management*, 212: 262–272.

Zhao Z, Qin X, Wang Z, Wang E. 2018. Performance of different cropping systems across precipitation gradient in North China Plain. *Agricultural and Forest Meteorology*, 258: 162–172.

Zheng C, Liu J, Cao G, Kendy E, Wang H, Jia Y. 2010. Can China cope with its water crisis? Perspectives from the North China Plain. *Groundwater*, 48(3): 350–354.

Zhong H, Sun L, Fischer G, Tian Z, van Velthuizen H, Liang Z. 2017. Mission impossible? Maintaining regional grain production level and recovering local groundwater table by cropping system adaptation across the North China Plain. *Agricultural Water Management*, 193: 1–12.

Part 2

Water-saving agriculture
One drop more crops

Chapter 3

Water-saving agriculture
Principles, regulation, and framework of practices

Huixiao Wang, Xiaohong Ren, Yanjun Shen, Yongqing Qi, and Changming Liu

3.1 Introduction

Water resource is the cornerstone for the survival and sustainable development of human society and the whole biosphere. In China, the south has much more water resources than the north, while the distribution of cultivated land is just reversed. The northern part of China is currently responsible for more than 65% of the national grain production, and an even higher proportion of fruits and vegetables production. However, the "more water for more grain" leads to severe over-exploitation of groundwater, the deterioration of the regional water cycle and ecological degradation (de Graaf et al., 2019; Liu et al., 2023). Therefore, sustainable development of agriculture and water resources requires advanced water-saving agriculture. Water-saving agriculture is a complex system of comprehensive utilization of water, soil, crop resources, and water conservancy engineering. Understanding the water transport between soil, crops, and the atmosphere is the theoretical basis of improving crop water-use efficiency (WUE) and developing water-saving agriculture (Shi et al., 1995; Liu, 2001; Wang et al., 2002).

Since the beginning of the 1990s, Liu Changming has led basic studies for agricultural water-saving practices in the North China Plain (NCP) and proposed the "Soil-Plant-Atmosphere Continuum (SPAC) interface agricultural water conservation and regulation theory" (Liu et al., 1999), which emphasizes that the process of water moving to the foliage from soil through the crop rhizomes and transpiring through the leaf stomata follows Ohm's law, the water transfer process in the SPAC system needs to break through a series of resistance, which are mainly concentrated at the different interfaces of SPAC, such as soil-root interface, soil-air interface, leaf-air interface. By taking measures to increase the resistance at the interfaces, water transport velocity and fluxes can be effectively reduced, thus achieving the purpose of saving water. Based on the theory of interfacial water conservation and regulation, the water-saving agriculture framework and its corresponding biological, agronomic, and engineering measures have been developed.

This chapter introduces the concept and application of the SPAC system and WUE, discusses the process of agricultural water consumption based on water cycle, and constructs a comprehensive framework of water-saving agricultural system at the field scale.

3.2 Water transferring in SPAC

3.2.1 SPAC basic theory and research progress

SPAC refers to using "water potential" to quantitatively study the change of energy level of each link in the SPAC system and calculate the energy of water movement (Philip, 1966). It links soil, plants, atmosphere, and the interface processes among them together to form a continuous, systematic, and dynamic system (Xu et al., 2013). Subsequently, Elfving et al. (1972) and Kaufmann et al. (1975) studied the process of water transport in the SPAC system, and investigated the influences of temperature and other environmental factors on water potential, which has important guiding significance for crop irrigation. The research on modern farmland moisture takes Soil-Vegetation-Atmosphere Transfer (SVAT) as the focus (Gong et al., 2002; Wu et al., 2014). With the deepening of small-scale studies, simulation of water and heat transfer in the SPAC system for different crops represented by Soil-Wheat-Atmosphere Continuum (SWAC) has become an important basis for water-saving agriculture (Yang et al., 1999). In subsequent studies on SPAC, the isotope technology, mathematical model, and other methods are applied to quantify and simulate the processes of water transport at the SPAC interfaces (Cienciala et al., 1994; Bariac et al., 1990; Saighi et al., 1998). The emergence of SPAC points out the direction of microscopic research on water cycle and strengthens interdisciplinary research on hydrology. Water potential, a unified energy relationship, makes the study of water movement and energy transformation processes in the SPAC system clearer and more convenient (Li, 2014; Liu et al., 1999).

As early as the 1980s, China had begun to carry out in-depth research on the measurement and estimation of water and heat fluxes in the SPAC system. After more than ten years of field observations and experiments, the researchers represented by Kang Shaozhong (1990, 1991, 1992) and his team explained the water movement in SPAC from the aspects of energy transport and water conversion, the mechanical-energy relationship of water transport, water absorption by plant roots, water transport in plants, and computer simulation. In subsequent studies, the groundwater was gradually paid attention to, and was included in the SPAC system. The concept of groundwater SPAC (GSPAC) system was proposed (Liu, 1997). It adds soil-groundwater interface and fully considers the unity and coordination of water and heat transfer between groundwater and SPAC (Shen, 1992; Liu, 1997). Then, Liu (2009) proposed the "five water" transformation theory based on the water cycle principle, which provided a theoretical basis for the extension of water movement theory to the mesoscale. In recent decades, research on SPAC mainly discussed the influence of soil water and groundwater on water consumption of crops, or used stable isotope method to reveal the process of water cycle and the mechanism of plant water use. The ultimate purpose of these studies is to improve the WUE of crops and develop water-saving agriculture.

3.2.2 Principles of water cycle and conversion in the SPAC system

The research on the water transport in the SPAC system is based on the principles of water cycle. Water transport in SPAC system (as in Figure 3.1) includes the processes of water absorption by roots, transport from root to leaf through stem, vaporization and transpiration in stomata, diffusion and transport in atmosphere through turbulence, and water evaporation from soil to air directly. These processes control the water cycle and the microclimatic environment of plant growth and have an important influence on the primary productivity (Liu et al., 1999).

The water transport process in the SPAC system mainly involves three interfaces: soil-atmosphere interface, soil-root interface, and plant-atmosphere (canopy-atmosphere) interface. The water fluxes of the three interfaces determine the water dissipation structure (transpiration and evaporation) of the farmland. By regulating the water transport at the soil-atmosphere interface, the water consumption of non-productive soil evaporation can be reduced. Regulating root water uptake can improve the utilization efficiency of soil water. By regulating water transport at the

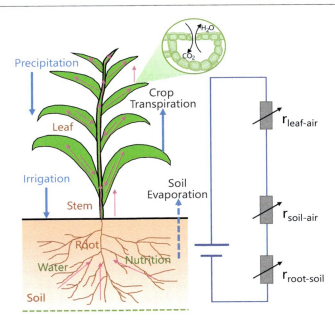

Figure 3.1 Schematic diagram of the water movement process and interface resistances in Soil-Plant-Atmosphere Continuum system.

plant-atmosphere interface, to increase stomatal resistance and reduce the intensity of water dissipation in photosynthesis process, crop transpiration can be reduced. The combined use of them ultimately achieves the purpose of improving the WUE of the entire farmland ecosystem.

On the basis of SPAC and GSPAC system principles, Liu Changming (2009) proposed the conversion of "five water", considering the water utilization of farmland crops and the basic application needs of water-saving agriculture. Farmland water operation and storage system includes five systems: atmosphere, soil, plants, surface and underground rock, known as the "five water" system. The corresponding "five water" are meteoric water, surface water, soil water, plant water, and groundwater. Driven by heat, gravity, and molecular forces, water circulates continuously in each system, including the main phase and state changes.

3.2.2.1 Water transport at the soil-root interface

In the SPAC system, water transport is always from high water potential to low water potential without energy input. Soil-root interface is an important subsystem in the SPAC system, and is the portal for roots to absorb water and nutrients and maintain plant growth. The flux at this interface is water absorption by roots, which has an important impact on crop growth and yield (Wang and Zhang, 2019). Sheng et al. (2005) showed that soil fertility and water condition determined the influence of soil resistance on root water absorption based on an one-dimensional soil hydrodynamic model. Liu (2013) also found that the soil water and fertilizer as well as the varieties significantly affected the water absorption capacity of plant roots. Soil fertility is the direct factor that causes the water potential difference between soil and root. Water redistribution is related to soil moisture and texture, and the physiological and ecological properties of plants such as water storage and conductivity (Wang and Zhang, 2019). Effective regulation of soil-root interface water can enhance root water absorption capacity and improve utilization efficiency of soil water.

3.2.2.2 Water transport at the plant-atmosphere interface

Plants are the central link in the SPAC system, which regulate water absorption and loss. There are three main processes involved in the exchange of matter and energy at the plant-atmosphere interface, namely photosynthesis, respiration and transpiration. The process of water transport from plants to atmosphere is mainly driven by root pressure and transpiration pull. Transpiration causes the water transport from soil-root-xylem-leaf tissue-stomata, which is a process of water dissipation. The water dissipation from leaves reflects the gas metabolism function of plants and the quantitative relationship between plant growth and water use, which is the physiological basis for improving WUE in the field. The coordination of photosynthesis and transpiration to obtain the most photosynthates with the least water consumption is important to improve the transpiration efficiency of leaves and the WUE of farmland. Many studies have shown that under the condition of full irrigation, WUE is not high due to the limited efficiency of photosynthetic production process, and mild water deficit can improve leaf photosynthesis and WUE (Sadok and Sinclair, 2011). For crops as a whole, the optimization of irrigation system will not only reduce leakage due to superfluous water supply, but also avoid water deficit due to too little water supply, which is of great significance for the NCP.

3.2.2.3 Water transport at the soil-atmosphere interface

The water flux at the soil-atmosphere interface mainly refers to infiltration and evaporation. Soil infiltration is an important part of water cycle and transformation, and the source of root water absorption and plant transpiration. Soil texture, moisture, and porosity affect the infiltration process to varying degrees. Adjusting the infiltration process by agricultural measures can effectively improve WUE. According to previous studies, soil evaporation accounts for 20%–40% of the total evapotranspiration of farmland, and reducing soil evaporation plays an important role in improving WUE of farmland (Wang et al., 1997). Optimizing the crop planting method can reduce soil evaporation by altering soil surface radiation, consequently improving WUE and saving water.

3.2.3 Simulation of water process in SPAC system

3.2.3.1 SPAC system moisture dynamic simulation method

At present, extensive studies have been carried out on the process of water transfer in the SPAC system. The results show that the dynamic change of water in the SPAC system has a certain scale effect, which can be divided into soil scale, farmland scale, and regional scale. The simulation of water dynamics can be divided into deterministic and stochastic method (Xu et al., 2013).

1. Deterministic Method: Deterministic method simulates water dynamics based on the physical law followed by water transport and transformation. It mainly includes conceptual model (water balance model), mechanical model (SPAC water transfer model, SPAC water and heat transfer model) (Shang, 2004). Water balance model is mainly applied at the field scale.

SPAC water transfer model is a soil hydrodynamic model established on the basis of Darcy's Law and continuous equation, which simultaneously considers soil evaporation, crop transpiration, and root water absorption processes. SPAC water and heat transfer model integrates theories of soil hydrodynamics, micrometeorology, and plant physiology to establish latent and sensible heat exchange between soil, canopy, and atmosphere. It considers soil water and heat transfer, evaporation and transpiration as a whole, and describes their dynamic changes.

2. Stochastic Method: The stochastic method includes mathematical statistics model, stochastic water balance model, and stochastic soil hydrodynamics model. The mathematical statistical models are based on the statistical relationship between soil moisture change and its main influencing factors or the change law of soil moisture itself, and mainly include statistical regression model, time series analysis model, regression index model, and artificial neural network model. The stochastic water balance and soil hydrodynamics models consider the temporal randomness and spatial variability of the inputs (precipitation, transpiration, etc.) and parameters (soil characteristics, etc.) of the water balance and soil hydrodynamics models.

3.2.3.2 SPAC commonly used comprehensive model of water transfer

With the deepening of the research on the water transfer process system, some comprehensive models were used to conduct systematic and complete simulation analysis. Table 3.1 shows several commonly used comprehensive models of the SPAC system, and briefly introduces the application fields, advantages, and disadvantages of the models.

Table 3.1 Common integrated model of Soil-Plant-Atmosphere Continuum system

Models	Main research objects and application areas	Main advantages	Main disadvantages
Waves (Zhang et al., 1996, 2016; Tian et al., 2017)	Simulate the dynamic transport process of water, energy and solutes; suitable for studying the hydrological and ecological responses caused by land management and climate changes	Dynamically simulate surface energy balance and distribution, vegetation growth and carbon distribution, vertical distribution of soil water, transpiration and other processes; easy to obtain data; adaptable widely; and able to use for long-term simulation if enough meteorological data are available	Integrated soil, canopy-atmosphere with a consistent level of process detail; ignore the differences between the subsystems of the model

(Continued)

Table 3.1 (Continued) Common integrated model of Soil-Plant-Atmosphere Continuum system

Models	Main research objects and application areas	Main advantages	Main disadvantages
Coup-Model (He et al., 2018)	Initially applied for simulating forest soil conditions, but now can be used for general illustration of water and heat processes in any soils with vegetation covering	Adaptable widely because a large part of the models are individual mechanistic ones; and flexible in time step	Limited in simulation because of the underlying assumptions of the model
Eco-Hat (Liu et al., 2009)	Realize the distributed regional eco-hydrological simulation including water and energy cycles; material transfer and transformation have been realized; provide scientific analysis tools for ecological hydrological process research, ecological benefit evaluation, and ecological design	Combine with RS and GIS, realizing the inverse combination of remote-sensing data with earth surface parameters; and designed modularly	Not widely used because the model source code is not open
Hydrus-1D (Jiménez-Martínez et al., 2009; Tan et al., 2015)	Used for simulating the distribution, temporal and spatial variation, and migration laws of water, solutes and energy in soils, and analyzing the practical problems of field fertilization, farmland irrigation, and environmental pollution	Adaptable widely; flexible in input and output; reliable result and able to use the exited databases	Preprocess the time step and some parameters, one-dimensional simulation has limitations
SWAP (van Dam et al., 2008)	Used for simulating water movement, solute migration, and heat transfer and crop growth at field scale, widely used to direct farmland irrigation	Practical and effective in irrigation water management and ecological environment protection	Difficult to describe the spatial variability of soil water-salt dynamics directly, due to the inherent one-dimensional structure and spatial limitations

SWAP model simulates the dynamic transport of water, solutes and heat, the crop growth and yield in the SPAC system. The upper and lower boundaries of this model are located above the canopy and groundwater saturation layer, respectively, and the dynamic changes of the natural environment have been widely used in arid and semi-arid regions. For example, Fan et al. (2007) used SWAP model to reveal the processes of plant water consumption and soil water movement under drought conditions in the wind-water erosion ecotone area. Yuan et al. (2014) applied SWAP model to simulate soil water and salt movement as well as yield under the growth condition of seed maize, and provided a theoretical basis for the establishment of salt water inadequate irrigation system. Liu et al. (2022) applied Hydrus-1D model to evaluate soil water and salt transport in response to varied rainfall events and hydrological years under brackish water irrigation in the NCP. Hydrus-2D model was used to simulate the distribution of soil water content under drip irrigation (Shan et al., 2019).

3.3 Crop WUE

Irrigation is a key means to ensure the yield and quality of crop production. However, in water-scarce areas, it is not enough to consider only the irrigation water use coefficient. Although advanced irrigation systems are visible to reduce water loss in addition to normal crop growth, the seemingly saved water is only transferred. Thereby the concept of WUE of crops is proposed. From the perspective of organisms, the WUE of crops refers to the amount of assimilation obtained per unit of water consumed by the crop, as distinct from irrigation WUE and precipitation use efficiency. Improving the WUE of crops is the ultimate goal of adopting engineering and non-engineering water-saving measures (Wang and Liu, 2000; Zhang et al., 2010).

3.3.1 Progress in research on crop WUE

Crop WUE reflects the energy conversion efficiency of crop production process, and is also one of the comprehensive indicators to evaluate the suitability of crop growth under water deficit. Research on WUE at the leaf level began in the 1960s (Wang and Liu, 2000). Montenith (1977) described crop yield in terms of the ability and amount of light intercepted by foliage and found that light, temperature, soil physical condition, and water supply are the main constraints on crop production efficiency. Farquhar and Richards (1984) suggested that carbon isotope analysis may be a useful tool in selection for improved WUE in breeding programs for C3 species. Dinar (1993) distinguished between potential and actual WUE measures, and discussed technology as well as greater emphasis on educational, extension activities, incentive, and price policies may also improve actual WUE levels. In recent years, the WUE-related models were used to explore the improvement of WUE from multiple perspectives such as irrigation management and genetics.

Early studies on WUE in China focused on arid and semi-arid regions, especially since the 1990s when water shortage had intensified and efficient use of water had received more attention. Crop WUE has different definitions in hydrology and crop physiology, while in most studies crop WUE is considered from the plant physiology perspective (Wang and Liu, 2000). Shan et al. (1991) proposed that the improvement of plant WUE is one of the main solutions to the key problem of developing water-saving agriculture. Studies on WUE involve three levels: leaf, plant, and population levels. Leaf level WUE can reveal the intrinsic water consumption mechanism of plants and provide a scientific basis for the rational water supply of vegetation, which is very useful for the development of water-saving agriculture in arid areas. Cao et al.

(2009) defined the leaf level WUE in detail and proposed the instantaneous WUE (WUEt) and the internal WUE (WUEi). Population level WUE can be divided into canopy and farmland levels (Jiang et al., 2019). Research on WUE in China ranges from microscopic leaf scale to macroscopic farmland and even ecosystem scale, and from various tillage practices before to during crop sowing. These studies provide strong evidence that developing water-saving agriculture and improving crop WUE are key, especially in arid and semi-arid areas.

Wang and Liu (2000) also systematically reviewed the concept of WUE as early as 1998 and defined it from two aspects of hydrology and crop physiology. In terms of hydrology, it mainly includes the following three aspects: (1) in pure hydrology, WUE refers to the ratio of regional productive water consumption to potentially available water; (2) in irrigation research, WUE can be defined as the increase in soil moisture of the root zone after irrigation as a proportion of the total water supply in the irrigated area; (3) total irrigation efficiency is composed of three components: water delivery efficiency, agricultural canal utilization efficiency, and field utilization efficiency. WUE in the hydrological sense is the ultimate goal of water saving in the scope of irrigation engineering and technology and contains items such as canal WUE, field WUE, and irrigation WUE. It includes research on regional water balance, on-farm water redistribution, diversion engineering and water deployment, canal impermeability, water transmission engineering, and new irrigation technologies. The hydrological study of crop WUE is an area of interest to engineers in irrigation and soil conservation, while research workers in physiology, agronomy, and meteorology are concerned with the concept, significance, and study of WUE physiologically. Physiological WUE is actually the WUE of a crop, which is a measure of crop yield in relation to water use. It is usually characterized by water consumption coefficient and WUE. The water consumption coefficient (Kw) is the amount of water consumed by a crop per unit of yield produced, often expressed as a multiple of the yield.

3.3.2 Calculation of crop WUE

3.3.2.1 Leaf level WUE

Leaf level WUE, also known as transpiration efficiency, is defined as the amount of organic matter formed by photosynthesis when a unit of water is dissipated by transpiration from leaves. It depends on the ratio of photosynthetic rate to transpiration rate and is the basic efficiency of water consumption by plants to form dry matter.

The net photosynthetic rate per unit leaf area is:

$$P_n = \frac{\Delta C}{r'_b + r'_s + r'_m} = \frac{C_a - \Gamma}{r'_b + r'_s + r'_m} \tag{3.1}$$

where P_n is the net photosynthetic rate of leaves; C_a is the air CO_2 concentration; Γ is the CO_2 compensation point (CO_2 concentration in the cytoplasm); ΔC is the difference between the air CO_2 concentration and the CO_2 compensation point; r'_b and r'_s are the boundary layer impedance and stomatal impedance of CO_2 diffusion into blades, respectively; r'_m is the chloroplast impedance to CO_2 diffusion into the cell chloroplast.

The transpiration rate per unit leaf area is:

$$T = \frac{\Delta H_2O}{r_b + r_s} = \frac{(\rho\varepsilon/P)(e_{ls} - e_a)}{r_b + r_s} \tag{3.2}$$

where T is the transpiration rate of the leaf; ΔH_2O is the difference between the water vapor concentration in the intercellular space and the atmospheric water vapor concentration; ρ and P are the air density and atmospheric pressure, respectively; ε is the ratio of the molar masses of water vapor and air; $e_{ls} - e_a$ is the leaf and air water vapor pressure gradient; r_b and r_s are the boundary layer and stomatal impedance of water vapor diffusion, respectively.

Then the WUE at the leaf level (WUE_l) is:

$$WUE_l = \frac{P_n}{T} = \frac{(C_a - \Gamma)(r_b + r_s)}{(\rho\varepsilon/P)(e_{ls} - e_a)(r_b' + r_s' + r_m')} \tag{3.3}$$

Based on the theory of molecular diffusion, the $r_s' = 1.56 r_s$ and $r_b' = 1.34 r_b$, the WUE_l can be further simplified.

3.3.2.2 WUE at the population level

The water-use efficiency of the crop (WUE_c) as a whole is the ratio of net CO_2 assimilation to transpiration by the crop population, that is, the ratio of the population CO_2 flux to the water vapor flux transpired by the crop.

$$WUE_c = F_c / T \tag{3.4}$$

where F_c is the crop community CO_2 flux; T is the water vapor flux transpired by the crop.

Field-scale CO_2 and water fluxes (ET) are determined using the Bowen ratio-energy balance method. The ET is calculated as:

$$ET = \frac{R_n - G}{(1 + \beta)\lambda} \tag{3.5}$$

where R_n is the net radiation; G is the soil heat flux; λ is the latent heat of vaporization; and β is the Bowen ratio, which can be calculated from the following equation:

$$\beta = \frac{C_p \rho \partial T_p}{\lambda \rho_a \partial W} \tag{3.6}$$

where C_p is specific heat at constant pressure; ρ is air density; ρ_a is the density of dry air; T_p is the mean potential temperature; and W is the air humidity (mass mixing ratio).

The CO_2 flux (F_c) is calculated as:

$$F_c = \frac{R_n - G}{C_p(\gamma + 1)} \frac{\partial C}{\partial T_e} \tag{3.7}$$

where γ is the ratio of the average density of water vapor to dry air; C is the CO_2 mass mixing ratio concentration; T_e is the effective temperature, which can be calculated from the following equation:

$$T_e = T_p + \frac{\lambda}{C_p} \frac{W}{\gamma+1} \tag{3.8}$$

3.3.2.3 WUE at yield level

Yield level WUE (WUE$_y$) is defined as yield per unit of water consumption. Yield can be expressed as net production or economic yield, where economic yield is closer to the actual agricultural production; water consumption takes into account the ineffective evaporation from the soil surface and is more practical for water conservation.

$$\text{WUE}_y = Y/WU \tag{3.9}$$

where Y is crop yield, which can be expressed as total biomass (Yb) or economic yield (Ye). WU is crop water use. According to water use, yield level WUE can be divided into three types, one is evapotranspiration efficiency obtained by crop water consumption, that is, ET; the second is irrigation WUE calculated by irrigation water (I); and the third is precipitation use efficiency derived by natural precipitation (P). One way to improve yield level WUE is to reduce ET without reducing crop yield, and another is to increase crop photosynthetic efficiency without increasing crop ET.

3.3.3 Influencing factors of crop WUE

3.3.3.1 Types and varieties of crops

As early as the beginning of the 20th century, Briggs and Shantz (1913, 1917) had started the work on plant WUE and found that there were some differences in WUE between different crops (Zhang, 2002). Zhang et al. (1989) compared the WUE of six rainfed crops on the Loess Plateau in China, of which the WUE of millet was higher than that of maize and wheat by 32% and 46%, respectively. Morgan et al. (1991) showed that the leaf intrinsic WUE of 15 winter wheat varied from 34.9 to 46.9 µmol/mol, and significantly differed by varieties. Zhang and Shan (1997) reported that nine wheat varieties differed by more than 40% in WUE of flag leaf. In the NCP, the increase in yield and WUE of maize was due to a significant increase in harvest index, which was related to the introduction of modern hybrid maize; overall, improvement of maize varieties contributed to 65% of the increase in WUE for the past 40 years (Zhang et al., 2021).

Considering water use varies considerably among different varieties of the same crop, and the high WUE characteristics of crops are heritable, regional WUE can be improved through variety selection in water-scarce areas.

3.3.3.2 Influencing factors of photosynthesis

Crop WUE depends on photosynthetic and transpiration rates. Light intensity indirectly affects crop WUE through influencing photosynthetic rate. However, the WUE will reach a stable constant or slightly decrease when the light intensity exceeds a certain threshold. Increasing CO_2 concentration can promote the photosynthetic rate and reduce the transpiration rate by regulating the stomatal conductance, then improve WUE, which is evidenced by an elevated CO_2 experiment of wheat (Liao and Wang, 2002).

Air humidity mainly affects transpiration rate, while having little effect on photosynthetic rate. Under low wind speed, transpiration rate decreases with increasing air humidity, and WUE increases with an exponential curve. Shi et al. (1994) found that leaf WUE of wheat decreased rapidly with the increase in vapor pressure deficit. Therefore, in actual production, sprinkler irrigation can be used to increase air humidity and reduce vapor pressure deficit to improve WUE.

Stomata are the channels for gas exchange between plants and the atmosphere, directly influence transpiration and photosynthesis, and inevitably affect crop WUE. When light and temperature conditions are suitable, stomatal conductance increases (stomatal resistance decreases), photosynthetic rate rises, leaf transpiration intensifies, and changes in WUE depend on the relative speed of the changes in transpiration and photosynthetic rates. It has been found that the effect of stomatal conductance on photosynthetic rate can be greater than the effect on transpiration rate (Xu, 1998). Thus, WUE can be improved by reducing leaf stomatal conductance.

3.3.3.3 Fertility

Fertilizer is an important environmental factor that affects crop growth and WUE. Zhang and Li (1996) found that under nutrient-deficient conditions, plant WUE could be significantly improved by applying fertilizer. It was also found that increasing soil fertility could reduce the effect of water deficit on photosynthetic rate under certain conditions, and finally improve WUE (Barati et al., 2015). Scientific and reasonable fertilization to achieve water regulation by fertilizer, root promotion by water, and drought resistance by root can improve the WUE of crops and promote the full utilization of limited water resources.

3.3.3.4 Soil moisture

Crop growth requires water uptake from soil through the roots to ensure the supply of physiological and ecological water requirements. It has been extensively shown that water stress decreases photosynthetic and transpiration rates, stomatal conductance, and intercellular CO_2 concentration, which eventually lead to yield loss. The crop WUE increases with the increase in water supply below a certain threshold of soil moisture, while it decreases when the soil moisture exceeds the threshold. Kang et al. (1998) found that a moderate water deficit at the seedling stage of maize facilitates the uptake of more water and nutrients at a later stage. Although the water deficit during the growing period reduces the yield of individual plant, the overall yield remains higher through dense planting, achieving efficient water use. A similar finding also exists in the response of wheat to soil moisture (Zhao et al., 2020). In water-scarce areas, controlled alternative irrigation, deficit-regulating irrigation, and other irrigation measures provide new directions for the development of water-saving agriculture.

3.4 Water-saving agriculture, a systematic framework

Water-saving agriculture refers to the rational development and utilization of water resources; the use of a variety of technical means to improve the effectiveness of water use within the agricultural model represents the integrated development and utilization of water, soil, and crop resources, which embody systematic engineering. Based on the pilot works in China's water-saving agricultural researches conducted by Liu et al. (1999), Shan (2003), and Kang et al. (1997, 1999) and the authors' experimental studies at the NCP, we proposed a comprehensive and systematic framework of water-saving agriculture (Shen et al., 2023). Water-saving

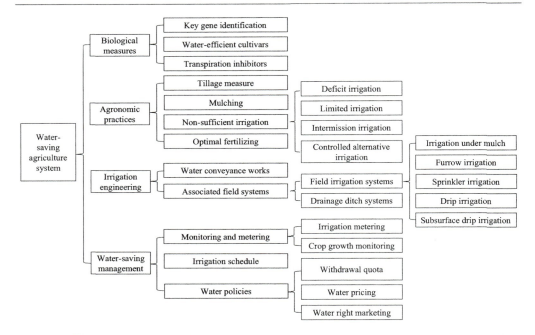

Figure 3.2 The comprehensive framework of water-saving agriculture system. (Adapted from Shen et al., 2023.)

agriculture includes four main groups of measures: biological, agronomic, engineering, and management.

The conceptual framework of integrated agricultural water-saving technical measures and research system in water-scarce areas is shown in Figure 3.2. Biological water-saving measures mainly include identification and utilization of high water-efficiency genes in crop breeding, cultivation of water-saving varieties, etc. Agronomic water-saving practices include conservation tillage, surface mulching, deficit-regulating irrigation and the controlled alternative irrigation and other non-sufficient irrigation, and fertilization system. Engineering water-saving measures include water transmission and allocation (e.g., canal seepage control works, pipeline transmission) and field irrigation system (e.g., water-saving irrigation such as furrow, sprinkler, and drip irrigation equipment). Water-saving management measures mainly include irrigation schedule optimization, water metering, monitoring and irrigation forecasting, and intelligent irrigation management technology. These measures can improve WUE in crop production and reduce irrigation water consumption individually and in interaction with each other.

3.4.1 Biological water-saving measures and technologies

The tolerance and adaptation of crops to water scarcity have changed as a result of climatic conditions and the environment, mainly including changes in the water absorption capacity of roots, the stomatal structure, and the efficiency of water vapor exchange during photosynthesis under drought conditions. The drought tolerance of plants is mainly characterized by their ability to regulate the osmotic potential of cells. During drought, cells can change their osmotic potential by increasing soluble substances, thus avoiding dehydration and maintaining a certain photosynthetic capacity. Drought tolerance varies significantly among crops, and millet, sorghum, and wheat are considered more drought-tolerant crops in arid and semi-arid agricultural areas.

It is a challenge to develop the drought-resistant, drought-tolerant, water-saving varieties with a higher WUE through genetic engineering. Modern water-saving breeding is based on the adaptability of crops to drought environment and aims at increasing crop yield and WUE through biological functions. The responses of different crops or varieties to water stress differ significantly and can result in difference in WUE by 30% among varieties. The WUE can be considered as a heritable trait in terms of yield and drought resistance, hence breeding new cultivars of drought tolerance or resistance could achieve the aim of water saving. A comparative study of the WUE of 59 wheat varieties in Hebei Province found that the average WUE under rainfed conditions was 1.45 kg/m^3, with the maximum value (1.77 kg/m^3) being 2.2 times the minimum value (0.81 kg/m^3), and that the average WUE under conventional irrigation was 1.70 kg/m^3, with the maximum value (2.09 kg/m^3) being 1.7 times the minimum value (1.22 kg/m^3) (Qiao et al., 2023). In the NCP, the increase in yield and WUE of maize was due to a significant increase in harvest index, which was related to the introduction of modern hybrid maize; overall, improvement of maize varieties contributed to 65% of the WUE increase, and WUE has increased from 1.35 to 2.36 kg/m^3 during the past four decades under full irrigation condition (Zhang et al., 2021).

Transpiration inhibitor is a drought-resistant chemical for crops with xanthate as the main ingredient. It regulates plant stomatal opening, reduces water transpiration, promotes root development, enhances root vigor, and increases chlorophyll content, thus promoting crop growth and increasing crop yields. The use of transpiration inhibitors is usually a drought mitigation measure rather than a conventional crop production measure.

3.4.2 Agronomic water-saving practices

Water-saving agronomic practices include mulching techniques to reduce soil evaporation, tillage techniques to maintain soil moisture, irrigation water control measures to improve crop WUE, and fertilization methods to synergistically promote water and fertilizer efficiency.

Reducing soil evaporation is an important component of water-saving agriculture. Measures adopted in field water-saving practices generally include mulching and water conservation tillage, in which the effects of straw mulching on saving water and increasing yield were most evident. Crop straw mulching also can significantly reduce soil evaporation. Meanwhile, mulching significantly increases yields, WUE (yield per unit water), and NUE (yield per unit N) of wheat and maize by up to 60%, compared with no mulching (Qin et al., 2015). In the NCP, applying maize-straw as mulch in the following wheat is a cost-effective practice. Tillage is a traditional soil water conservation technique and is an important part of agricultural water saving. The difference in soil porosity between tilled and untilled soils is about 10%, and the difference in infiltration is about 20%. Deep tillage can enlarge soil water storage capacity. The soil with deep tillage could store over 90% of rainwater in the autumn, and the upper-layer soil water was increased by 10.2%–12.0% after summer tillage for two years.

It has been shown that balanced fertilizer application is an important way to increase field WUE. Crop stomatal regulation, water-holding capacity, membrane permeability, and photosynthesis are closely linked to nitrogen (N), phosphorus (P), and potassium (K) application. Stomatal conductance and transpiration rate at high and moderate N levels are lower than those at low N level under water stress. N application could increase the leaf net photosynthesis rate under high or moderate water stress, while no difference in the leaf net photosynthesis rate existed at severe water stress (Barati et al., 2015).

In China, surface irrigation accounts for 98%, and a number of advanced non-sufficient surface irrigation methods have already been developed. Such methods improve WUE by controlling the amount of water irrigated in the field, reducing unnecessary water leakage, and creating moderate water-deficit conditions. Border check irrigation of small plots, or basin irrigation, can reduce irrigation water volumes, but raise irrigation uniformity up to over 80%. This technique can also control deep drainage to prevent the rise of groundwater tables and soil salinization. Surge irrigation or intermittent irrigation has been found to greatly improve the efficiency of surface ditch or border irrigation by raising water conveyance. It supplies water intermittently to field irrigation ditches or plots at specified time intervals. It can save up to 10%–40% of water compared to ordinary gravity irrigation due to the reduction of deep drainage and tail water loss.

3.4.3 Engineering water-saving measures

Engineering and technical water-saving measures mainly include water conveyance structures, water-saving irrigation techniques and equipment. Delivering water by pipes can effectively reduce seepage and evaporation. At the end of 2020, the irrigation area using low-pressure pipes had increased to over 3 million ha in Hebei Province, water-saving surface irrigation has basically been realized in the agricultural area of plain region. The water conveyance coefficient of pipes is now over 0.9 and could save water by up to 50%, 15%, and 7% compared with an earth canal, a canal lined with stone, and a canal lined with concrete in irrigation districts using groundwater, respectively; and it could save around 40% in irrigation districts using surface water.

Sprinkler and drip irrigation not only reduce irrigation water amount and build a more reasonable soil moisture condition, but also greatly improve crop production efficiency and mechanization of other field management activities through automated operation. Sprinkler and drip irrigation are more commonly used in vegetable and fruit growing. Compared to surface drip irrigation, subsurface drip irrigation can reduce soil evaporation even more, and improve irrigation WUE to a greater extent (Yao, 2021). A four-year field experiment showed that subsurface drip irrigation technology significantly improved the utilization of irrigation water. Compared with traditional surface irrigation, evapotranspiration can be reduced by 88 mm in wheat season and by 60 mm in maize season; the annual field water consumption could be reduced by 1,480 m^3/ha without yield loss observed (Umair et al., 2019; Yao, 2021).

3.4.4 Water-saving management

In general, better management can help make an efficient water-use system, which could ensure successful implementation of other water-saving measures and obtain good effectiveness. The water-shortage problem could not be solved solely based on water-saving technologies. Economic and administrative measures are also necessary to ensure sound water management. The water resources fee could provide incentives to use groundwater more efficiently (Kemper, 2007).

In 2016, Hebei Province was selected as the pilot province for the fee to tax transformation in water resources management, and great positive effects of groundwater pumping reductions were observed immediately. From 2017, the fee to tax reform has been expanded to nine provinces in northern and western China. The central government of China has been issuing regulations to promote the development of a water rights system over the past two decades. The first two important regulations were issued in 2005, that is, 'Some Opinions on Water Rights Transfer' and 'Establishing a Framework of Water Rights System'. In 2014, the government launched formal pilot projects in seven provinces to further accelerate the development of water rights means. However, it is more difficult to establish a water rights system to promote rights transfer among irrigation water users.

With the implementation of the strictest water resource management legislation, the measurement and monitoring of agricultural irrigation water use have become one of the main elements of groundwater resource control, and a means to the total amount of irrigation pumping and irrigation WUE. Due to poor facilities, high labor cost, and complex situations in reality, it is hard to charge irrigated groundwater in the field. However, since most tube wells include electricity metering, and the major operation cost of a tube well is the electricity cost, it is common to collect groundwater irrigation fees based on electricity use. Charging groundwater irrigation fees according to electricity use was treated as a proxy approach to a volumetric irrigation fee in practice in Hebei Plain (Wang et al., 2020; Li et al., 2022). In recent years, measures such as water right reforms and water-saving incentives piloted in Cheng'an, Taocheng, and Yuanshi counties in Hebei Province have achieved better water-saving effects and formed good experiences (Ma et al., 2019).

3.5 Concluding remarks

In most areas of the world, water shortage is caused by overuse of water in the agricultural sector. As the biggest water consumptive user, improving the agricultural WUE is crucial for solving regional water shortage and securing food production. As an important grain and food producer in China, the NCP uses only 6% of the country's water resources to support 18% of the country's arable land and produce 23% of the country's grain production.

Based on the biophysical processes of water transfer in SPAC, a systematic approach to water saving can be formulated, encompassing regulatory mechanisms at interfaces and its integrated measures, culminating in a thorough water-saving agriculture framework. Under this framework, efforts both in research and administrative communities should be endeavored together to promote the crop WUE or water productivity, and manage water use more efficiently for achieving a water-saving agriculture and sustainable future.

"Water-saving agriculture" is a complex system involving biologic, agronomic, and hydraulic engineering techniques in the integrated exploitation of water, soil, and crop resources. Only when water-saving agriculture is considered as an integrated system, comprehensive water-saving measures can be properly evaluated and applied. Therefore, water-saving agriculture should be the first priority with water transfer projects to follow. Water-saving agriculture will provide a good basis on which the magnitude of water transfer can be optimized to minimize cost and to maximize the benefit of the projects. Sociologically, water saving would mitigate the conflicts between the water-rich and the water-scarce regions. In other words, water saving is not only essential for solving water-shortage problems but also socio-economically and eco-environmentally beneficial. Improving the management of water resources is an extremely important issue. There are urgent needs to establish advance-with-the-times groundwater management legislation and associated supporting technologies, especially the intelligent systems comprised of the devices in monitoring, sensing, metering, as well as decision-making tools.

References

Barati V, Ghadiri H, Zand-Parsa S, Karimian N. 2015. Nitrogen and water use efficiencies and yield response of barley cultivars under different irrigation and nitrogen regimes in a semi-arid Mediterranean climate. *Archives of Agronomy and Soil Science*, 61(1): 15–32.

Bariac T, Jusserand C, Mariotti A. 1990. Evolution spatio-temporelle de la composition isotopique de l'eau dans le continuum sol-plante-atmosphère. *Geochimica et Cosmochimica Acta*, 54(2): 413–424.

Briggs L J, Shantz H L. 1913. *The water requirement of plants*. Washington, United States: US Government Printing Office.

Briggs L J, Shantz H L. 1917. The water requirement of plants as influenced by environment. *Proceedings of the Second Pan American Scientific Congress*, 3: 95–107.

Cao S, Feng Q, Si J, Chang Z, Xi H, Zhuomacuo. 2009. Summary on research methods of water use efficiency in plant. *Journal of Desert Research*, 29(5): 853–858 (in Chinese).

Cienciala E, Eckersten H, Lindronth A, Hallgren J E. 1994. Simulated and measured water uptake by Picea abies under non-limiting soil water conditions. *Agricultural and Forest Meteorology*, 71(1–2): 147–164.

de Graaf I E M, Gleeson T, van Beek L P H, Sutanudjaja E H, Bierkens, M F P. 2019. Environmental flow limits to global groundwater pumping. *Nature*, 574(7776): 90–94.

Dinar A. 1993. Economic factors and opportunities as determinants of water use efficiency in agriculture. *Irrigation Science*, 14(2): 47–52.

Elfving D C, Hall A E, Kaufmann M R. 1972. Interpreting leaf water potential measurements with a model of the soil-plant-atmosphere continuum. *Physiologia Plantarum*, 27(2): 161–168.

Fan J, Wang Q, Shao M. 2007. Numerical modeling of the soil-water dynamics in water-wind erosion crisscross region on the loess plateau. *Advances in Water Science*, 18(5): 683–688 (in Chinese).

Farquhar G D, Richards R A. 1984. Isotopic composition of plant carbon correlates with water-use efficiency of wheat genotypes. *Australian Journal of Plant Physiology*, 11(6): 539–552.

Gong L, Pan X, Chang S, Zhang X. 2002. Achievements of study on soil-plant-atmosphere Continuum and discussion its application in arid zone. *Environmental Protection of Xinjiang*, 24(2): 1–4 (in Chinese).

He H, Meyer A, Jansson P E, Svensson M, Rütting T, Klemedtsson L. 2018. Simulating ectomycorrhiza in boreal forests: implementing ectomycorrhizal fungi model MYCOFON in CoupModel (v5), *Geoscientific Model Development*, 11(2): 725–751.

Jiang H, Zhang Y, Ren X, Yao J, Shen Y. 2019. A review of progress in research and scaling-up methods of crop water use efficiency. *Chinese Journal of Eco-Agriculture*, 27(1): 50–59 (in Chinese).

Jiménez-Martínez J, Skaggs T H, van Genuchten M T, Candela L. 2009. A root zone modelling approach to estimating groundwater recharge from irrigated areas. *Journal of Hydrology*, 367(1–2): 138–149

Kang S, Liu X, Gao X, Xiong Y. 1992. Computer simulation of water transport in soil-plant-atmosphere continuum. *Journal of Hydraulic Engineering*, 3(10): 1–12 (in Chinese).

Kang S, Liu X, Wang Z. 1991. Relations between leaf water potential, stomatal resistance and transpiration rate of winter wheat, and environmental factors. *Journal of Irrigation and Drainage*, 10(3): 1–6 (in Chinese).

Kang S, Ma X. 1999. Thoughts of several problems on developing the economizer of water agriculture in our nation. *Journal of China Agricultural Resources and Regional Planning*, 20(2): 30–32 (in Chinese).

Kang S, Shi W, Hu X, Liang Y. 1998. Effects of regulated deficit irrigation on physiological indices and water use efficiency of maize. *Transactions of the Chinese Society of Agricultural Engineering*, 4: 88–93 (in Chinese).

Kang S, Xiong Y, Wang Z. 1990. Distribution of hydraulic resistance and water potential in soil-plant-atmosphere continuum. *Journal of Hydraulic Engineering*, 7: 1–9 (in Chinese).

Kang S, Zhang J, Liang Z, Hu X, Cai H. 1997. The controlled alternative irrigation: a new approach for water saving regulation in farmland. *Agricultural Research in the Arid Areas*, 15(1): 1–6 (in Chinese).

Kaufmann M R. 1975. Leaf water stress in Engelmann spruce: influence of the root and shoot environments. *Plant Physiology*, 56(6): 841–844.

Kemper K. 2007. Instruments and institutions for groundwater management. In Giordano M, Villholth K G (Eds.), *The agricultural groundwater revolution: Opportunities and threats for development*. Cambridge, MA: CABI. pp: 153–172.

Li F, Tao P, Qi Y, Li H, Wang H, Wang N, Pei H, Zhang X. 2022. Factors influencing electricity-to-water conversion metering method for irrigation water consumption in Hebei Plain. *Chinese Journal of Eco-Agriculture*, 30(12): 1993–2001 (in Chinese).

Li Y. 2014. *Water transport process and mechanism of soil-plant-atmosphere continuum of typical forest ecosystems in Beijing mountainous area*. Beijing, China: Beijing Forestry University (in Chinese).

Liao J, Wang G. 2002. Effects of drought, CO_2 concentration and temperature increasing on photosynthesis rate, evapotranspiration, and water use efficiency of spring wheat. *Chinese Journal of Applied Ecology*, 13(5): 547–550 (in Chinese).

Liu B, Wang S, Liu X, Sun H. 2022. Evaluating soil water and salt transport in response to varied rainfall events and hydrological years under brackish water irrigation in the North China Plain. *Geoderma*, 422: 115954.

Liu C. 1997. Study on interface processes of water cycle in soil plant atmosphere continuum. *Acta Geographica Sinica*, 52(4): 366–373 (in Chinese).

Liu C. 2001. Water-saving priority, water demand control, open source and reduce expenditure. *Water Resources Development Research*, 1: 3–4 (in Chinese).

Liu C, Sun R. 1999. Ecological aspects of water cycle: advances in soil-vegetation-atmosphere of energy and water fluxes. *Advance in Water Science*, 10(3): 251–259 (in Chinese).

Liu C, Wang H. 1999. *Water processes and water-saving regulation at soil-crop-atmosphere interface*. Beijing, China: Science Press (in Chinese).

Liu C, Yang S, Wen Z, Wang L, Wang Y, Li X, Shen H. 2009. Development and application of distributed eco-hydrological model EcoHAT system. *Science in China (Series E: Technological Sciences)*, 39(6): 1112–1121 (in Chinese).

Liu M, Guo Y, Zhang X, Shen Y-J, Zhang Y, Pei H, Min L, Wang S, Shen Y. 2023. China's Black Soil Granary is increasingly facing extreme hydrological drought threats. *Science Bulletin*, 68(5): 481–484.

Liu X. 2013. *Regulation mechanism of water uptake by maize roots*. Ph D dissertation. Beijing, China: University of Chinese Academy of Sciences (in Chinese).

Ma S, Sun M, Fu Y, Yao Z, Li S. 2019. Methods and practice use of water right confirmation in Hebei province. *South-to-North Water Transfers and Water Science & Technology*, 17(4): 94–103 (In Chinese)

Madani E M, Jansson P E, Babelon I, 2018. Differences in water balance between grassland and forest watersheds using long-term data, derived using the CoupModel. *Hydrology Research*, 49(1): 72–89.

Montenith J L. 1977. Climate and the efficiency of crop production in Britain. *Philosophical transactions of the Royal Society of London*, 281(980): 277–294.

Morgan J A, Lecain D R. 1991. Leaf gas exchange and related leaf traits among 15 winter wheat genotypes. *Crop Science*, 31(2): 443–448.

Philip J R. 1966. Plant water relations: some physical aspects. *Annual Review of Plant Physiology*, 17: 245–268.

Qiao W, Yang H, Yang H, Li Y, Ma Y, Zhong Y, Qiao Y, Dong B. 2023. Screening of winter wheat varieties for drought-resistant, water-saving and high yield and water use efficiency. *Journal of Triticeae Crops*, 43(10): 1254–1266 (In Chinese)

Qin W, Hu C, Oene O. 2015. Soil mulching significantly enhances yields and water and nitrogen use efficiencies of maize and wheat: a meta-analysis. *Scientific Reports*, 5: 16210.

Sadok W, Sinclair T R. 2011. Crops yield increase under water-limited conditions: review of recent physiological advances for soybean genetic improvement. *Advances in Agronomy*, 113: 313–337.

Saighi M, Moyne C. 1998. A new simplified model of heat and mass transfer in the soil-plant-atmosphere system. *International Journal of Heat and Mass Transfer*, 41(11): 1459–1471.

Shan G, Sun Y, Zhou H, Lammers P S, Grantz D A, Xue X, Wang Z. 2019. A horizontal mobile dielectric sensor to assess dynamic soil water content and flows: direct measurements under drip irrigation compared with HYDRUS-2D model simulation. *Biosystems Engineering*, 179: 13–21.

Shan L. 2003. Issues of science and technology on water saving agricultural development in China. *Agricultural Research in the Arid Areas*, 21(1): 1–5 (in Chinese).

Shan L, Xu M. 1991. Water-saving agriculture and its physio-ecological bases. *Chinese Journal of Applied Ecology*, 2(1): 70–76 (in Chinese).

Shang S. 2004. Advances in soil moisture simulation and forecasting models. *Journal of Shenyang Agricultural University*, 35(5~6): 455–458 (in Chinese).

Shen Y, Qi Y, Luo J, Zhang Y, Liu C. 2023. The combined pathway to sustainable agricultural water saving and water resources management: an integrated geographical perspective. *Acta Geographica Sinica*, 78(7): 1718–1730 (in Chinese).

Sheng Y, Zhao C, Jia H. 2005. Effects of water and fertilizer coupling on soil moisture movement in corn fields. *Arid Land Geography*, 28(6): 811–817 (in Chinese).

Shen Z. 1992. *Water resources scientific experiment and research-atmospheric, surface, soil and ground water interactions*. Beijing, China: Science and Technology of China Press (in Chinese).

Shi J, Wang T, Li L. 1994. Experimental study and mathematical simulation of water use efficiency in wheat leaves affected by some environmental factors. *Acta Botanica Sinica*, 36(12): 940–946 (in Chinese).

Shi Y, Liu C, Gong Y. 1995. *Advances in basic research of water-saving agriculture application*. Beijing, China: China Agriculture Press (in Chinese).

Tan X, Shao D, Gu W, Liu H. 2015. Field analysis of water and nitrogen fate in lowland paddy fields under different water managements using HYDRUS-1D. *Agricultural Water Management*, 150: 67–80.

Tian F, Feng X, Zhang L, Fu B, Wang, S, Lv Y, Wang P. 2017. Effects of revegetation on soil moisture under different precipitation gradients in the Loess Plateau, China. *Hydrology Research*, 48(5): 1378–1390.

Umair M, Hussain T, Jiang H B, Jiang H, Ahmad A, Yao J, Qi Y, Zhang Y, Min L, Shen Y. 2019. Water-saving potential of subsurface drip irrigation for winter wheat. *Sustainability*, 11(10): 2978.

van Dam J C, Groenendijk P, Hendriks R F A, Kroes J G. 2008. Advances of modeling water flow in variably saturated soils with SWAP. *Vadose Zone Journal*, 7(2): 640–653.

Wang H, Liu C. 1997. Evapotranspiration, soil evaporation and water efficient use. *Acta Geographica Sinica*, 52(5): 65–72 (in Chinese).

Wang H, Liu C. 2000. Advances in crop water use efficiency research. *Advances in Water Science*, 11(1): 99–104 (in Chinese).

Wang H, Liu C, Liu Z. 2002. Water-saving agriculture in China: an overview. *Advances in Agronomy*, 75: 135–171.

Wang L, Kinzelbach W, Yao H, Steiner J, Wang H. 2020. How to meter agricultural pumping at numerous small-scale wells? An indirect monitoring method using electric energy as proxy. *Water*, 12(9): 2477.

Wang L, Zhang H. 2019. Analysis of water circulation and transformation process in SPAC system. *Water Resources Planning and Design*, 8: 41–44 (in Chinese).

Wu S, Mo F, Zhou H, Asfa B, Zhao H, Deng H, Chen Y, Xiong Y, Zhang H. 2014. Research progress of soil-water dynamics model applied in SPAC system. *Agricultural Research in the Arid Areas*, 32(1): 100–109 (in Chinese).

Xu L, Xu J, Dong L, Feng W, Jiang J. 2013. Research advance in process and modeling of water transfer in soil-plant-atmosphere continuum. *Agricultural Research in the Arid Areas*, 31(1): 242–248 (in Chinese).

Xu Y. 1998. Research progress of crop water use efficiency. *Shanxi Agricultural Sciences*, 4: 14–18 (in Chinese).

Yang J, Li B, Li Y, Ma R. 1999. Preliminary studies on groundwater effects on SPAC system in shallow groundwater field. *Journal of Hydraulic Engineering*, 7: 28–33 (in Chinese).

Yao J. 2021. *Study on water use efficiency of typical farmland in North China Plain under subsurface drip irrigation*. Beijing, China: University of Chinese Academy of Sciences (In Chinese)

Yuan C, Feng S, Jiang J, Huo Z, Ji Q, Qi Y. 2014. Simulation of water-salt transport by SWAP model under deficit irrigation with saline water. *Transactions of the Chinese Society of Agricultural Engineering*, 30(20): 72–82 (in Chinese).

Zhang K, Kimball J S, Running S W. 2016. A review of remote sensing based actual evapotranspiration estimation. *Wiley Interdisciplinary Reviews-Water*, 3(6): 834–853.

Zhang L, Dawes W R, Hatton T J, 1996. Modelling hydrologic processes using a biophysically based model application of WAVES to FIFE and HAPEX MOBILHY. *Journal of Hydrology*, 185(1–4): 147–169.

Zhang S, Li Y. 1996. Study on effects of fertilizing on crop yield and its mechanism to raise water use efficiency. *Research of Soil and Water Conservation*, 3(1): 185–191 (in Chinese).

Zhang S, Shan L. 2002. Research progress on water use efficiency of plant. *Agricultural Research in the Arid Areas*, 20(4): 1–5 (in Chinese).

Zhang X, Chen S, Sun H, Wang Y, Shao L. 2010. Water use efficiency and associated traits in winter wheat cultivars in the North China Plain. *Agricultural Water Management*, 97(8): 1117–1125.

Zhang X, Uwimpaye F, Yan Z, Shao L, Chen S, Sun H, Liu X. 2021. Water productivity improvement in summer maize: a case study in the North China Plain from 1980 to 2019. *Agricultural Water Management*, 247: 106728.

Zhang X, Xu Y, Shan L. 1989. Requirement to water on different crops under different dryland conditions. *Acta Ecologica Sinica*, 9(1): 97–98 (in Chinese).

Zhang Z, Shan L. 1997. Comparative study on water use efficiency of wheat flag leaves. *Chinese Science Bulletin*, 42(17): 1876–1881 (in Chinese).

Zhao W, Liu L, Shen Q, Yang J, Han X, Tian F, Wu J. 2020. Effects of water stress on photosynthesis, yield, and water use efficiency in winter wheat. *Water*, 12(8): 2127.

Chapter 4

Water budgets, ET partitioning, and implications to water saving

Yucui Zhang, Yanjun Shen, and Fan Liu

4.1 Introduction: energy/water budgets and evapotranspiration partitioning

Water and energy are the most basic elements of life, and are also the most active and influential factors in the ecosystem. All life activities on earth are restricted by water and heat conditions. In the earth system, water is not only an integral part of the material cycle, but also a medium for the circulation of other biological materials. Driven by energy, water continuously circulates between organisms, environments, or organisms and the environment, maintaining the existence, succession, and development of ecosystems.

Australian hydrologist Philip proposed the "Soil-Plant-Atmosphere Continuum" (SPAC) (Philip, 1966), which takes soil, plants, and atmosphere as a physical continuum to explore the relationship between soil water and plants. Water reaches the root epidermis of the plant through the soil, enters the root system, passes through the plant stem, reaches the leaf, and then diffuses through the leaf stomata to the air, and finally participates in the turbulent exchange of the atmosphere, forming a unified dynamic mutual feedback continuous system. In this system, the growth of vegetation is determined by the light, heat, and moisture conditions. Simultaneously, it is accompanied by energy partitioning between latent and sensible heat through hydrological processes such as canopy transpiration. The process in turn affects the water cycle of the basin. Water and energy processes are closely coupled with biophysical and physiological processes of the vegetation.

After reaching the earth's surface, solar radiation is often converted into various forms of energy, such as latent heat of evaporation or condensation, sensible heat of turbulence. For crops, grasslands, and bare land, the heat storage of surface to vegetation canopy and the energy consumed by photosynthesis are very small and can generally be ignored, so the energy budget of land surface can be expressed as follows according to the law of conservation of energy:

$$Rn = H + LE + G \tag{4.1}$$

where Rn is the net radiation (MJ/m^2), H is the sensible heat flux, LE is the latent heat flux, G is the soil heat flux (MJ/m^2). The energy balance equation is applicable to various time and spatial scales.

Net radiation is the sum of incident short-wave and long-wave radiation minus the outgoing short-wave and long-wave radiation:

$$Rn = S_{down} - S_{up} + L_{down} - L_{up} \tag{4.2}$$

where S_{down} is the short-wave radiation that reaches the ground from the sky, S_{up} is the short-wave radiation reflected from the ground, L_{down} is the downward long-wave radiation, and L_{up} is the upward long-wave radiation.

Latent heat flux can be calculated by:

$$LE = \lambda ET \qquad (4.3)$$

in which ET is the evapotranspiration (mm), an important part in water balance; λ is the latent heat of vaporization (MJ/kg). Generally, it can be calculated by air temperature (equals $2.501 - 2.361 \times 10^{-3} \times T$ (°C)). Therefore, evapotranspiration is the key factor linking water balance and energy balance.

Land surface water balance refers to the difference between the amount of water received and the water spent in any period in a selected area, equal to the change in water storage during that time frame. Water balance in an irrigated farmland can be expressed in the following form:

$$ET = P + I - R - D - \Delta W \qquad (4.4)$$

where ET is evapotranspiration (mm), P is precipitation (mm); I is irrigation application (mm), R is surface runoff (mm), D is soil water drainage (mm), and ΔW is change in soil water storage in the measured soil depth range during the growing stage. Generally, surface runoff was ignored in the flat farmland.

The partitioning of evapotranspiration into evaporation and transpiration determines the water consumption for productive use (Transpiration) and non-productive use (Evaporation). Quantifying ET partitioning can help understand the water-use structure and take measures to enlarge the proportion of transpiration in water consumption, then enhance water-use efficiency (WUE).

The proportion of T to ET varies across different ecosystems in different regions. Research on the grasslands ecosystem in Inner Mongolia revealed T/ET contributions of 24% ± 13% for shrubs and 20% ± 4% for grasses because of the difference in ecohydrological connectivity and leaf development (Wang et al., 2018). Experiments conducted in Japan showed the T/ET was 0.74 for rice from May to September (Wei et al., 2018). In northwestern China, transpiration contributed 87% to the total ET during the growing season (Wen et al., 2016) and in the North China Plain (NCP), it was 83% at the late grain filling stage and 60% at wax ripeness stage of the wheat (Zhang et al., 2011). In a comprehensive review focused on the ET partition by Xiao et al. (2018), 33 studies published from 1997 to 2018 were analysed, incorporating the stable isotope two-source mixing model encompassing about 18 types of ecosystems. However, there remains a gap in ET partition research for the entire growth periods of crucial grain crops, especially in regions with serious water shortage and large population to support.

Regardless of transpiration or evaporation, the water used by crop ecosystem comes from soil water. Determining the main water uptake depths of crop and soil evaporation depth could provide vital information of how the soil water is used by crops and is lost "uselessly" through soil evaporation. This knowledge is fundamental to develop technologies/measures to prevent soil evaporation loss and shift soil water being exhausted only from crop root-xylem-leaf systems to join the process of photosynthesis as much as possible. This is considered as one path to explore the potential of water saving in agricultural practices.

We will present the studies conducted to quantify ET and its partitioning, through energy and water budget measurement and analysis, for different agro-ecosystems in NCP. Soil water-use/consumption mechanisms of wheat-maize double cropping system is also introduced in detail including the methods of combining isotopic and micro-meteorologic approaches, the exhaustion structure and its implications, and the subsurface drip irrigation (SSDI) technology developed according to these findings.

4.2 Energy/water budgets and WUE of major agro-ecosystems in Ncp

Water cycle and energy balance are inseparable. Net radiation of the ecosystem is the energy source that drives the temperature change, facilitating sensible and latent heat exchanges on the underlying surface of vegetation. The transmission direction and amount of soil heat flux directly affect the level and variation of soil temperature. The latent heat is the main component of energy balance, and the corresponding ET is the main way of crop water consumption. When the sensible heat and latent heat are different or the surface temperature is not uniform, it will cause thermal advection in the horizontal direction, so that the wetter cropland can obtain advection heat from the surrounding drier areas and increase the water consumption of crops. This further affects the energy balance and water cycle of cropland. Therefore, a comprehensive understanding of energy balance is crucial for evaluating the efficacy of various water-saving measures.

4.2.1 Field setting of energy, water, and carbon fluxes observations

Energy and water balances at land surface are significantly affected by environmental factors such as atmospheric temperature, humidity, radiation, precipitation, as well as the changes in soil moisture and irrigation. These factors, combined with the different structures and characteristics of the underlying surface, create distinct energy and water budgets, shaping the specific microclimate of farmland.

We conducted observations in four distinct sites representing different agro-ecosystems for grain, cotton, fruit orchard, and vegetable field ecosystem (Figure 4.1). These four agro-ecosystems represent the main planting types in NCP, which also hold important position and account for 23%–34% in China (Zhang et al., 2022). The distance between sites varied from 20 to 200 km. In situ observations were conducted over the four agro-ecosystems for 3–13 years (Table 4.1).

The meteorological factors, such as the annual radiation, average relative humidity, average temperature, and annual precipitation, were very similar at the four sites (Table 4.1). Our experiments obtained successive data during 2017–2018 for greenhouse vegetable systems and one ~9 m profiles were sampled in 2013 for the open-air vegetable ecosystem (Min et al., 2018). Soil physical properties, fertilizer application, and growth stages are shown in Table 4.1 and Figure 4.2.

4.2.2 Energy/water budgets of the major agro-ecosystems

The process of cropland energy budget is extremely important for crop/atmosphere interaction research and the formulation of farmland water management measures. Cropland energy flux directly affects temperature, water transport, vegetation development, and ecosystem productivity. Wheat and maize are the most important crops with the largest planted area in North China, so we took wheat and maize cropland as an example to analyse its energy budget.

Water budgets, ET partitioning, and implications to water saving 53

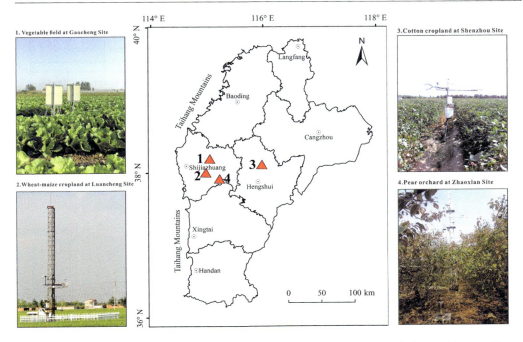

Figure 4.1 Locations of the four experimental sites for energy/water balance observation in NCP.

Table 4.1 Crop types, groundwater depths (GD), soil bulk density (SBD), water–N-fertilizer input, and meteorological conditions (soil moisture: θ, annual radiation: Ra, average daily relative humidity: RH, average daily temperature: Ta, annual precipitation: P) of the four sites

Crop types	Exp. periods	GD (m)	SBD (g/cm³)	N-input (kgN/ha)	θ (%)	Ra (MJ/m²)	RH (%)	Ta (°C)	P (mm)
Wheat-maize	Oct. 2007 to Sep. 2019	45	1.53	450–600	28	4,194	71	12	466
Pear trees	Aug. 2011 to Sep. 2019	50	1.32	540–800	30	4,617	73	13	475
Cotton	Oct. 2015 to Sep. 2019	50	1.43	260–280	26	~5,000	67	13	465
Vegetables	2013, 2017, and 2018	40	1.20	700–900	29	~4,300	75	13	464

Crop canopy receives net radiation with peaks ranging from 16 to 19 MJ/m²/d in June for the wheat-maize agro-ecosystem (Figure 4.3). Annual Rn ranges from 2,235 to 2,671 MJ/m²/year, with an average of 2,438 MJ/m²/year (Table 4.2). Seasonal Rn of wheat was 18–468 MJ/m² higher than that of maize season, mainly caused by less precipitation and longer growth periods during the wheat season. Temporal variations in soil heat flux (G) generally correspond to Rn. G could account for up to 4% of annual Rn. As the main component of the energy fluxes, the

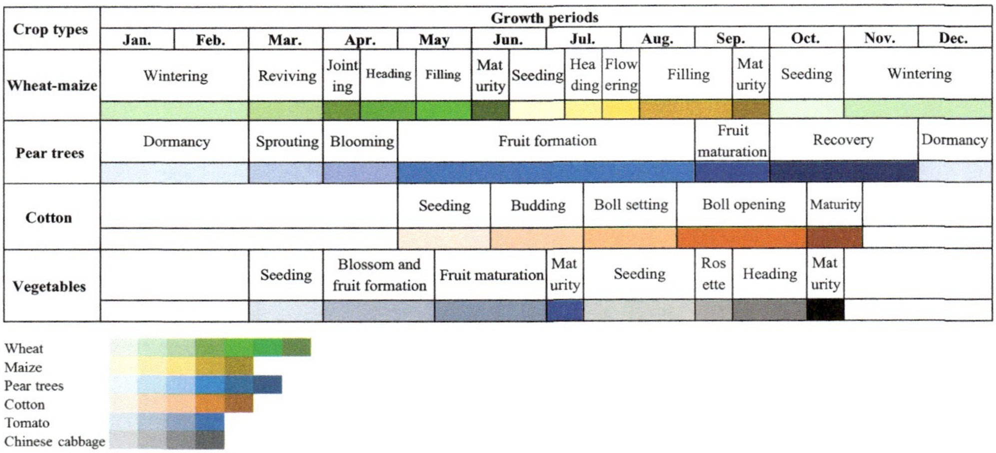

Crop types	Growth periods											
	Jan.	Feb.	Mar.	Apr.	May	Jun.	Jul.	Aug.	Sep.	Oct.	Nov.	Dec.
Wheat-maize	Wintering		Reviving	Jointing / Heading	Filling	Maturity / Seeding	Heading / Flowering	Filling	Maturity	Seeding	Wintering	
Pear trees	Dormancy		Sprouting	Blooming	Fruit formation				Fruit maturation	Recovery		Dormancy
Cotton						Seeding	Budding	Boll setting	Boll opening	Maturity		
Vegetables				Seeding	Blossom and fruit formation	Fruit maturation	Maturity	Seeding	Rosette	Heading	Maturity	

Wheat
Maize
Pear trees
Cotton
Tomato
Chinese cabbage

Figure 4.2 Growth stages of the four types of agro-ecosystems.

latent heat flux (LE) consumed 74% of the total net radiation. Seasonal LE of wheat season was 19%–53% higher than that in the maize season. The sensible heat flux (H) showed an almost opposite variation trend compared to LE. Peak daily H was recorded during crop rotation from winter wheat to summer maize in mid-June, when H can reach ~3–10 MJ/m²/d. The ratio of annual H/Rn ranged from 18% to 28%.

Table 4.2 also gives the annual means of energy balance components of pear orchard and cotton field agro-ecosystems. In the pear orchard, the annual Rn varied from 2,503 to 2,793 MJ/m²/year, which was higher than that of wheat and maize cropland. Rn allocated a higher proportion (76%–81%) to LE relative to H and G. The ranges of LE, H, and G were 1,961–2,122, 465–731, and -73—3 MJ/m²/year, with the corresponding mean values of 2,064, 561, and -18 MJ/m²/year, respectively.

Rn in cotton ecosystem was lower than that in the grain field and pear orchard, and fell in the range of 2,329–2,371 MJ/m²/year, of which the proportion of LE accounted for 56%–64%. The LE, H, and G changed from 1,332 to 1,479, 850 to 1,018, −1 to 20 MJ/m²/year, with the average of 1,433, 913, and 9 MJ/m²/year, respectively. Seasonally, the Rn and LE of the pear orchard and cotton ecosystems showed a similar bell-shaped curve, and the peaks occurred in summer months, while the H displayed a reversed pattern.

ET links water and energy budgets and directly determines the WUE. The water for ET was mainly sourced from soil layers, frequently exhausted and refilled by precipitation and irrigation. Through measuring soil water storage and its change, the evapotranspiration and deep drainage can be calculated. Figure 4.4 illustrated the field water balance for wheat-maize ecosystem in detail.

The annual ET of wheat-maize cropland was 714.2 mm, while the average monthly P was significantly different, varying from 0.6 to 135.5 mm (Figure 4.4a). The highest monthly water demand occurred in May, and the average monthly P was only 29.0 mm. Local precipitation is mainly concentrated in June to September, accounting for 77.4% of the whole year, with a peak of 135.5 mm in August. The soil water storage of 0–2 m layer fluctuated around 485.4 mm (Figure 4.4b). The lowest soil water storage was in early June, averaging of 459.5 mm, coinciding with the grain harvest stage of wheat. Low soil water storage continued until the wheat harvest in

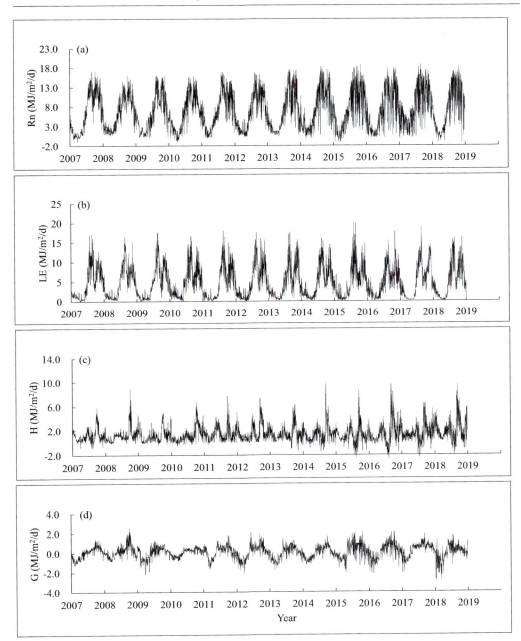

Figure 4.3 Panels (a) variations of daily net radiation (Rn), (b) latent heat flux (LE), (c) sensible heat flux (H), and (d) soil heat flux (G) at Luancheng Station during 2007–2018.

Table 4.2 Energy budgets of different agro-ecosystems

Types of the ecosystems	Energy balance items (MJ/m²)			
	Rn	LE	H	G
Cotton field	2356±19	1433±68	913±74	9±11
Maize cropland	1070±75	786±69	245±35	39±14
Wheat cropland	1368±101	1074±81	310±73	−17±29
Pear orchard	2607±82	2064±55	561±78	−18±28

Rn, net radiation; LE, latent heat flux; H, sensible heat flux; G, soil heat flux.

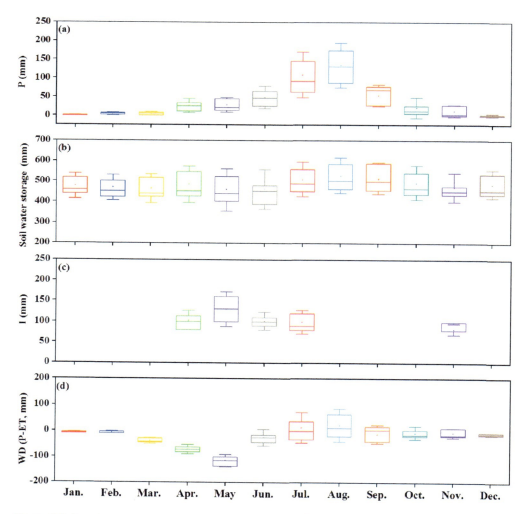

Figure 4.4 Panels (a) average monthly precipitation (P), (b) soil water storage, (c) irrigation (I), (d) water deficit (WD) of the winter wheat–summer maize cropland ecosystem during the observation period of October 2007 to September 2019.

Table 4.3 Water balance of different agro-ecosystems

Types of the ecosystems	Water balance items (mm)					
	ET	P	I	ΔW	D	P-ET
Cotton field	592 ± 25	465 ± 152	132 ± 110	−5 ± 52	11 ± 40	−127
Maize cropland	305 ± 17	325 ± 93	114 ± 39	57 ± 39	77 ± 65	20
Wheat cropland	409 ± 17	129 ± 52	249 ± 66	−56 ± 60	25 ± 12	−280
Pear orchard	786 ± 32	475 ± 143	391 ± 104	−1 ± 41	81 ± 81	−311
Vegetable field	997 ± 31	461 ± 0	829 ± 156	0	293 ± 128	−536

Notes: ET, evapotranspiration; P, precipitation; I, irrigation; ΔW, soil water storage; D, drainage; P-ET represents water deficit. The ΔW of vegetable field is a supposed value.

mid-June, which is the main reason for the negative ΔW. Soil water storage was highest in August, with an average of 530.1 mm and a maximum of 663.6 mm, mainly due to the concentrated precipitation.

Irrigation is generally applied two to four times with the amount of 80 mm in November, April to May (wintering, jointing, flowering, and early filling stages) in the wheat season (Figure 4.4c), and one to two times concentrated in June to July (pre-seeding and heading stages) during the maize season. Water deficit (WD) reflects the minimum demand of irrigation. Monthly WD showed the largest WD reaching 115.0 mm in May and the largest surplus in August (Figure 4.4d). This result was closely determined by the uneven distribution of rainfall and the water consumption of crop growth. Except for the water surplus in July to August, other months experienced deficit and need irrigation to replenish.

Water balance results for the other two main agro-ecosystems are shown in Table 4.3. The distribution of P observed at the four agro-ecosystem is similar. The annual irrigation of vegetable field is the highest, more than 800 mm, followed by orchards of 391 mm; cotton fields have the smallest irrigation frequency and amount (132 mm/year). Deep drainage for different ecosystems was similar to that for irrigation. Vegetables field consumes water at the most, nearly 1,000 mm/year, having a huge impact on groundwater depletion and pollution risk (through large amount of fertilizer input and deep drainage). The water expressed by the difference between precipitation and ET was also the highest (536 mm) in vegetable fields, and the lowest (127 mm) in cotton fields.

4.2.3 Net ecosystem carbon exchange and WUE of the agro-ecosystems

The net ecosystem carbon exchange (NEE) can reflect the biomass accumulation of the ecosystem over a period, which has close relationship with the ecosystem productivity. We can get the WUE of an ecosystem by the ratio of NEE and ET.

The net ecosystem productivity of typical cropland in NCP shows a "W"-shaped distribution throughout the year, corresponding to two peaks in the vigorous growing seasons of wheat and maize. Under the current planting system and management measures, the two peaks of daily net ecosystem productivity can reach 15 gC/m²/d (maize filling period in mid-August) and 13 gC/m²/d (wheat filling period in early May), respectively. NEE is closely related to the stage of crop growth and development, and it varies greatly in different months. The annual NEE of cropland is about 534 gC/m², of which the maize season is higher than the wheat season. The average NEE of wheat season can reach 259 gC/m², with a range from 120 to 400 gC/m² in different years. The average NEE of maize season can reach 275 gC/m², and the variation range in different years is 160–415 gC/m².

The peak of NEE of pear orchards appeared from May to September. NEE variation of pear orchard ecosystem is different from the cropland ecosystem which has two absorption peaks, and the value of the peak period changes slowly, which lasts longer than that of the farmland ecosystem. The difference in NEE between pear orchard ecosystem and cropland ecosystem is mainly caused by different phenological change characteristics. For the different growth stages of pear trees, NEE of the pear orchard during the flowering stage accounted for 4.5% of the total growth season, while fruit formation-expansion period and fruit maturation-harvest period accounted for 62.4% and 12.5% of the total, respectively (Figure 4.5). The annual average NEE could reach −667 gC/m².

The daily variation pattern of NEE in cotton fields is similar to that in pear orchards. The average annual NEE of cotton field ecosystems from 2015 to 2019 was −351 gC/m².

The WUE of different agricultural ecosystems varies greatly, and the changes with the crop growth process are also different. In this study, plant photosynthesis was much less than crop and soil respiration at the rotation or fallow periods, thus the ecosystems were in carbon release, WUE based on productivity had no practical significance. Therefore, we mainly focused on the changes in WUE during the growing season (April and May for wheat and mid July to September for maize from 2007 to 2019). Like the seasonal patterns of NEE of the wheat and maize

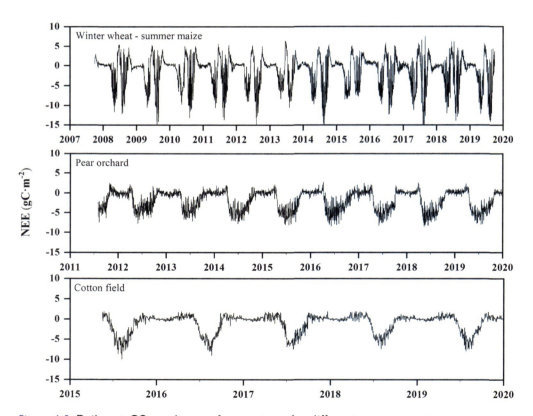

Figure 4.5 Daily net CO_2 exchange of ecosystems for different agro-ecosystems.

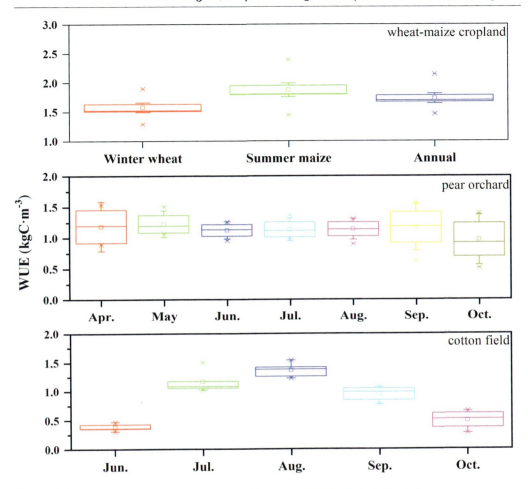

Figure 4.6 WUE of the major agro-ecosystems at seasonal or monthly scales.

agro-ecosystem, the WUE showed two peaks throughout the year (Figure 4.6). The pattern of two peaks was directly related to the cropping system of winter wheat and summer maize. The peak of WUE (2.5 kgC/m^3) occurred in late April to mid-May (early filling stage). Subsequently, at the flowering stage, wheat transitioned from vegetative growth to reproductive growth, the dry matter accumulation decreased significantly, and led to a decrease in WUE. The peak of WUE in maize season (3.6 kgC/m^3) appeared in mid-August (similar growing stage as wheat). The WUE of maize was higher than that of wheat due to the C4 photosynthesis path of maize. The mean productivity-based WUE of wheat and maize throughout their respective growing season was 1.6 and 1.9 kgC/m^3, respectively The coefficients of variation of WUE in wheat and maize seasons were 10% and 7%, respectively.

The seasonality of WUE in the pear orchard differed from that of wheat and maize agro-ecosystem. The vigorous growing period, May to October, did not show an obvious peak. The WUE varied from 0.5 to 2.5 kgC/m^3, a slightly higher value occurred in May and September, as ET was relatively small and NEE remained at a high level during this period. Although the carbon accumulation was relatively large from July to August, the water consumption during the same period was also high. The average WUE during the growing season of pear trees (from April to September) from 2011 to 2019 was 1.2 kgC/m^3, which was slightly lower than that of wheat and maize agro-ecosystem (1.7 kgC/m^3). Peak WUE (2.7 kgC/m^3) of cotton field occurred in late July or early August. The mean WUE at seedling, flower budding, flower and boll, and boll opening stages was 0.2, 0.8, 1.4, and 0.97 kgC/m^3, respectively. The mean WUE from June to October (0.88 kgC/m^3) was lower than that of pear orchard and wheat and maize agro-ecosystem.

4.3 ET partitioning and soil water-use mechanism

As the significant role of the winter wheat and summer maize cropland, detailed knowledge of water cycling within this agro-ecosystem is required in order to inform cultivation practices that can achieve better WUE. In particular, refining strategies to increase the ratio of transpiration to other soil water balance sink terms (e.g. evaporation) necessitate (1) ET partitioning; (2) distribution of the major layers of soil water for root uptake throughout the growing season; (3) characterization of the evaporation depth for the wheat-maize cropland. The water consumption structure is available through water flux measurements. The isotopic methods also can be used in the determination of water uptake or evaporation depth, based on the differences in isotopic compositions (d^{18}O and d^{2}H) of stem water and soil water. Water consumption patterns and mechanism of croplands under general field management will help to carry out targeted water consumption management of crops in different periods, formulate scientific and reasonable irrigation plans.

4.3.1 ET partitioning of different agro-ecosystems

The separation of ET components is an important basis for accurate estimation of WUE. Vegetation transpiration is the necessary water for crop growth, which is related to the type and biological characteristics of crops, while evaporation from soil surface is considered as non-productive consumption, which is related to atmospheric rainfall, irrigation, and soil water content.

Based on the hydrogen and oxygen stable isotope method, ET partitioning of wheat-maize field was estimated by employing Keeling plot (Yepez et al., 2003; Keeling, 1961) and Craig–Gordon model (Craig and Gordon, 1965). The calculation details can be found in Zhang et al. (2011). The ratio of E/ET ranged from 11.5% to 56.4% during the whole growing season (Figure 4.7). On a seasonal scale, evaporation accounted for 36.8% of ET in winter wheat growing season and 34.3% in summer maize growing season. The proportion was 35.3% for the whole wheat-maize season. These three values were 33.2%, 35.1% and 34.3%, correspondingly, measured by a weighing mini-lysimeter method. This result is consistent with the model simulation, which indicated that evaporation accounted for 40% of ET over the whole wheat-maize season (Umair et al., 2017).

Besides the isotope method, evaporation of wheat-maize cropland and cotton field was also measured by micro-lysimeter (Sun et al., 2004), to compare the water consumption structure of different agro-ecosystems. Meanwhile, the transpiration of pear orchard was measured by the needle sap flow meter (Thermal Dissipation Probe, TDP30-FLGS-TDP XM1000, Dynamax, USA). Figure 4.8 shows the monthly ET partitioning of the three agro-ecosystems.

The highest monthly T/ET was 0.82 for wheat and 0.77 for maize. Lower T/ET appeared in the crop rotation months (June and October) and winter months. The proportion of T/ET was 66.4% in the winter wheat season and 61.7% in the summer maize season, respectively. The annual average T/ET was 64.4% and T was 459.9 mm, which means E/ET was 35.6% for the

Water budgets, ET partitioning, and implications to water saving 61

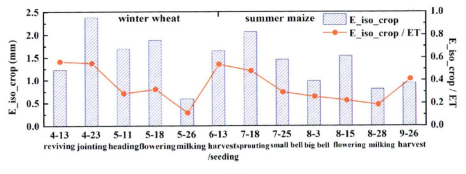

Figure 4.7 Evapotranspiration partition by isotope method (E_iso_crop: evaporation calculated by isotope method).

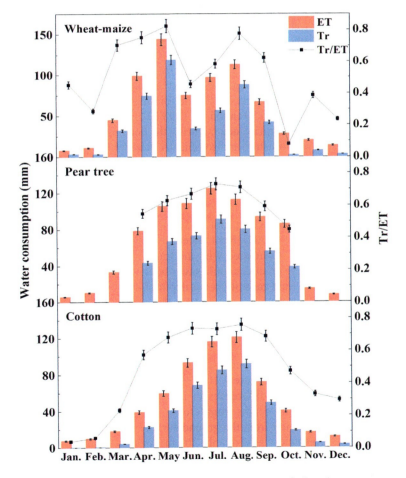

Figure 4.8 Monthly evapotranspiration (ET) and its partition of the three agro-ecosystems (Tr: Transpiration).

wheat-maize cropland ecosystem. This result was also similar with the isotope method and the model simulation.

The T/ET of pear orchard was 73.0%, much lower than that of wheat-maize ecosystem. Transpiration of the pear orchard ecosystem was 485.7 mm, with 25.8 mm higher than that of the wheat-maize cropland ecosystem, and the evaporation was 300.3 mm of the pear orchard.

Transpiration by the cotton field was 381.2 mm and accounted for 64.4% of ET. Transpiration in cotton field was 104.5 and 78.7 mm lower than that of pear orchard and wheat-maize cropland ecosystems, respectively.

4.3.2 Main root water uptake and evaporation depth

ET partitioning quantifies the water consumption for plant transpiration and soil evaporation. However, what is the source of water absorption by plant roots? Does the root system absorb the same amount of water from different soil layers? Do the roots in irrigated and rain-fed croplands utilize water in the same manner?

Root water uptake depths were examined using a comparison of concurrently measured $\delta^{18}O$ change in the soil profile, stem water oxygen isotopic composition, and soil water contents (Zhang et al., 2011). The depths of root water uptake are inferred from the depths where stem water and soil water isotopic compositions are identical (Wang et al., 2010), considering that there is no hydrogen and oxygen isotope fractionation effect during transpiration by most plants, including winter wheat (Lin and Sternberg, 1994; Yakir and Sternberg, 2000). The results suggest that root water uptake is the shallowest during the jointing stage (10 cm, DOY92, Figure 4.9). The main uptake layer is from 20 to 30 cm on the DOY138, in the milking stage, and at 20 cm on the DOY149 standing for the maturity stage. During the heading stage, there are three depths at which the stem water and soil moisture are isotopically equal (10 cm and between 20 and 40 cm; DOY116). But the results could be that the stem water was a mixture of soil water from different layers at different relative rates. Therefore, further analysis was undertaken using the IsoSource multiple-source isotopic mass balance model (Phillips and Gregg, 2003; Phillips et al., 2005).

The root water uptake occurs mainly above 40 cm during all growing stages of wheat. These results are consistent with winter wheat root densities that are much greater above 40 cm than within 40–100 cm (Zhang et al., 2004). Root water absorption rates of winter wheat (Zuo et al., 1998) also support this result. Compared with the traditional wetting depth of 100 cm, widespread implementation of these practices could save about 240 m^3/ha/year water if we assume the mean soil water content and field capacity at the depth of 40–100 cm was 0.30 and 0.34 m^3/m^3, respectively. This assumption does not account for the occasional flushing of the root zone to remove accumulated minerals.

Soil evaporation occurrence can be well revealed by the isotope enrichment process. To understand the variations in evaporation between bare soil and crop field, isotope compositions at the same depth during crop growing periods were averaged, and the isotope characteristics at different depths were compared (Figure 4.10; Zhang et al., 2022). Significant isotope enrichment (p-value is 0.02) and large differences in isotope compositions for bare and crop field soil water were observed for soil depths from 0 to 30 cm. These results indicated the impact of crops on evaporation distribution from the ground surface to 30 cm, and the evaporation depth for both bare soil and crop field was probably 30 cm. As shown in Figure 4.7, the mean daily evaporation intensity was 0.64 mm/d for the winter wheat season, and 1.27 mm/d for the maize season. The mean daily evaporation intensity of the maize season was twice that of the wheat season. Compared with our previous study, the evaporation depth for the winter wheat season

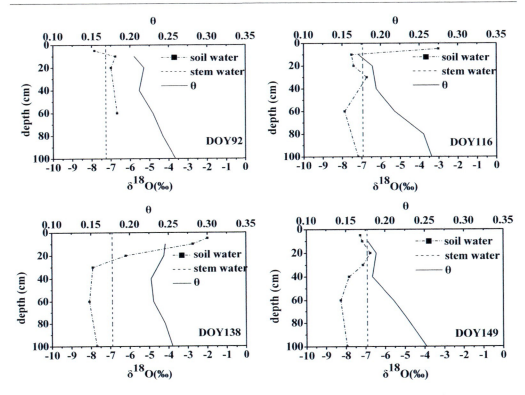

Figure 4.9 $\delta^{18}O$ signature of soil water from each depth and stem water including soil water content in different growth periods of winter wheat. (θ represents the soil water content; DOY indicates the day of year, 4 sub figures refer to different growing stages of wheat at Jointing (DOY92); Heading (DOY116); Milking (DOY138); and Maturity (DOY149), respectively.

obtained by comparing the isotope compositions of soil water in the rain-fed and well-watered treatments was 20 cm (Zhang et al., 2011). We inferred that stronger evaporation intensity increased the crop field soil evaporation depth from 20 cm during the winter wheat season to 30 cm during the summer maize season.

4.4 Exploring field water-saving potential through SSDI

Evaporation from the soil surface results in a considerable loss of moisture and has a direct impact on crop yield. As the most important crops in the NCP, water consumption of wheat and maize cropland will be reduced up to almost 40% in the form of soil evaporation. In addition, evaporation mainly occurred from the soil water above 20–30 cm depth and water for transpiration mainly from the 0–40 cm soil depth. Meanwhile, the depth of rotary tillage is generally shallower than 20 cm. These facts suggest that by altering the irrigation method and planned wetting depth to reduce the amount and adjust the position of application, we may achieve significant water saving through reduction of the soil evaporation loss. Then, we designed some experiments and related simulation studies to explore drip irrigation under soil surface and its water-saving potential.

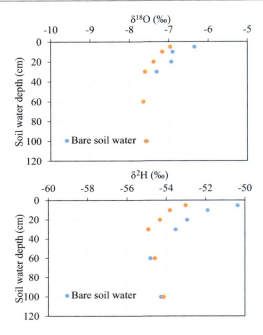

Figure 4.10 Isotope composition of soil moisture with different depths between bare soil water and crop field soil water treatments.

4.4.1 Determining water movement and loss of SSDI

We firstly measured the dynamics of soil wetting, water content, and evaporation under different treatments, then simulated these dynamics by the Hydrus model, and finally determined the optimal depth of dripper setting for SSDI. Five dripper depth (0, 15, 20, 25, and 30 cm) treatments were conducted in a glass box experiment (Figure 4.11). The soil wetting pattern is described by the wetting radius and depth away from the dripper (Figure 4.12; Umair, 2019). As the depth of the dripper increases, the time taken to reach the soil surface lengthens, and the horizontal migration rate slows down. The simulated soil wetting pattern by the Hydrus model is in good correspondence with the observed data. The measured and simulated distribution of soil water content under five depths of dripper treatments is shown in Figure 4.12. Centring around

Figure 4.11 Experimental boxes for irrigation and dynamics of soil wetting under different drip depths.

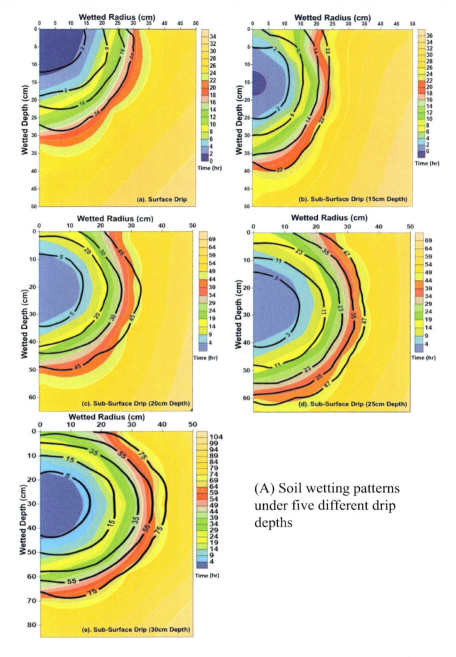

(A) Soil wetting patterns under five different drip depths

Figure 4.12 Soil wetting patterns (A) and the distribution of soil water content under five drip depths (B): (a) SDI, (b) SSDI: depth 15 cm, (c) SSDI: depth 20 cm, (d) SSDI: depth 25 cm, (e) SSDI: depth 30 cm at different application times (hr). Black bold contours represent observed wetting patterns (hr). Colour contours represent simulated wetting patterns (hr).

(Continued)

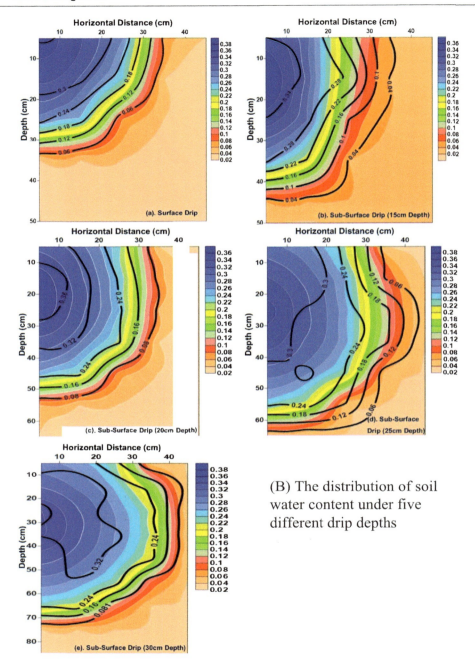

Figure 4.12 (Continued) Soil wetting patterns (A) and the distribution of soil water content under five drip depths (B): (a) SDI, (b) SSDI: depth 15 cm, (c) SSDI: depth 20 cm, (d) SSDI: depth 25 cm, (e) SSDI: depth 30 cm at different application times (hr). Black bold contours represent observed wetting patterns (hr). Colour contours represent simulated wetting patterns (hr).

4.4.2 Water-saving ability and water productivity of SSDI

We compared the yield and water balance under different irrigation methods in lysimeter, including the flood irrigation (FI), surface drip irrigation (SDI), and SSDI with a 25-cm dripper depth, and assessed the water-saving abilities of SDI and SSDI relative to FI.

The ET and yield for wheat and maize seasons under different irrigation methods are given in Figure 4.13 and Table 4.4. In the wheat season, the highest ET was found in FI and the lowest in SSDI (Figure 4.13). The SSDI and SDI treatments decreased ET by 80 and 40 mm, respectively, compared to FI. Grain yield increased by 5.68% in FI and 3.41% in SSDI relative to the yield in SDI (Table 4.4). In the maize season, the SSDI and SDI treatments reduced ET by 117 and 38 mm, respectively, compared to FI. The grain yield increased by 4.6% in SSDI and 4.47% in SDI relative to that in FI (Table 4.4). In summary, the SSDI saves water by limiting soil surface wetting and reducing ET, meanwhile, it can maintain the comparable grain yield relative to FI and SDI by keeping the wet condition of subsurface soil throughout the vegetative growth period.

4.4.3 ET and WUE of different treatments for SSDI

Three irrigation amounts (W_1, W_2, and W_3) and three drip tape spacing (D_1, D_2, and D_3) (for detailed differences see Table 4.5) under SSDI were conducted in the field trials, and the FI treatment was used as the reference. The ET, yield, and irrigation water-use efficiency (IWUE) under different treatments of wheat and maize are shown in Table 4.5. The irrigation treatments had significant effects on yield, ET, WUE, and IWUE for wheat. The yield of wheat in FI treatment

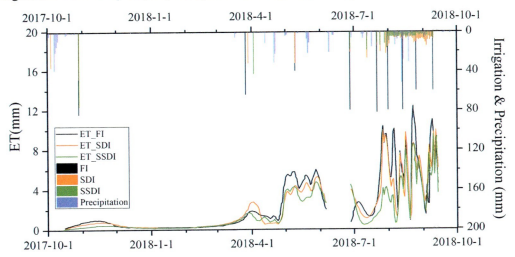

Figure 4.13 Comparison of ET for different irrigation methods during the wheat and maize growth period in 2017–2018. The lines show ET variations for different irrigation treatments, the bars for precipitation and irrigation amount. FI is flood irrigation; SDI, surface drip irrigation; SSDI, sub-surface drip irrigation.

Table 4.4 Yield components, grain yield, and aboveground biomass of winter wheat and summer maize season under three irrigation methods

Treatment of wheat	Number of spikes	Number of grains	Thousand-grain weight (g)	Yield (kg/ha)	Biomass (kg/ha)
FI	110	3806.0	32.4	4694.8	9610.8
SDI	107	3449.8	31.8	4428.0	8988.8
SSDI	117	3572.0	34.2	4584.8	9930.8

Treatment of maize	Number of spike rows	Spikes number	Hundred-grain weight (g)	Biomass (kg/ha)	Yield (kg/ha)
FI	16	29	38.96	12465.6	6609.2
SDI	16	33	32.85	13393.2	6918.4
SSDI	16	35	33.43	12292	6927.6

FI, flood irrigation; SDI, surface drip irrigation; SSDI, subsurface drip irrigation.

Table 4.5 Yield and water-use efficiency of wheat and maize under different subsurface drip irrigation treatments

Treatment of wheat	Number of grains per spike	Thousand-grain weight (g)	Yield (kg/ha)	ET (mm)	WUE (kg/m^3)	IWUE (kg/m^3)	Harvest index
W_1D_1	31.3	42.1	6,244	313.8	2.0	3.7	0.48
W_1D_2	32.1	40.9	6,673	307.5	2.2	4.0	0.51
W_1D_3	31.9	42.0	6,438	320.3	2.0	3.8	0.47
W_2D_1	31.2	40.9	5,944	304.8	1.9	4.2	0.46
W_2D_2	29.5	41.2	6,253	293.6	2.1	4.5	0.48
W_2D_3	29.2	41.6	5,987	305.4	2.0	4.3	0.47
W_3D_1	34.0	41.9	6,680	266.1	2.5	5.8	0.51
W_3D_2	31.1	40.6	6,667	268.8	2.5	5.8	0.51
W_3D_3	34.4	41.5	6,542	271.5	2.4	5.7	0.51
FI	31.9	39.4	6,989	365.9	1.9	3.2	0.50

Treatment of maize	Number of rows per spike	Number of grains per spike	Hundred-grains weight (g)	Yield (kg/ha)	ET (mm)	WUE (kg/m^3)	IWUE (kg/m^3)
W_1D_1	14.2	39.5	26.9	8,930	293.7	3.0	14.4
W_1D_2	13.5	36.7	29.0	8,885	299.2	3.0	14.3
W_1D_3	13.8	38.3	28.1	9,112	260.3	3.5	14.7
W_2D_1	13.9	38.6	27.2	8,250	278.7	3.0	17.2
W_2D_2	14.0	35.9	28.1	8,477	279.8	3.0	17.7
W_2D_3	14.2	36.0	28.7	8,749	280.4	3.1	18.2
W_3D_1	14.4	35.8	28.3	8,024	273.9	2.9	22.9
W_3D_2	13.7	34.8	26.6	8,170	272.3	3.0	23.3
W_3D_3	14.5	35.2	27.1	8,320	270.1	3.1	23.8
FI	13.7	39.0	32.7	9,642	324.6	2.9	12.0

Different lowercase letters mean significant differences at 0.05 level among different treatments. W_1, W_2, and W_3 indicate the irrigation amount is 167, 140, and 115 mm, respectively. D_1, D_2, and D_3 represent the belt spacing of 60, 80, and 100 cm, respectively.

ET, evapotranspiration; FI, flood irrigation; WUE, water-use efficiency; IWUE, irrigation water-use efficiency.

was higher than SSDI treatment, and decreased with the irrigation amount under the three SSDI treatments. For different drip tape spacing of SSDI, there was no significant differences of ET with the average value of 294.8 mm for wheat season. Therefore, the most appropriate spacing for drip irrigation tape was 100 cm according to the construction cost. However, ET of wheat diverged significantly among W_1, W_2, and W_3 treatments, with the values of 313.9, 301.3, and 268.8 mm, respectively. And ET of maize showed the similar results. The SSDI significantly increased the WUE and IWUE under low irrigation conditions. The WUE and IWUE of FI treatment was the lowest, and SSDI improved the IWUE by 21.6%–81.6%. Compared with FI treatment, the IWUE was improved by 20.1%, 46.8%, and 93.7% for W_1, W_2, and W_3 treatments, respectively. In summary, the SSDI technology significantly reduced ET and improved the WUE and IWUE. As a result, the average total water-saving ability was 1,500 m^3/ha of the cropland, which was ~100 mm for winter wheat season and ~50 mm for summer maize season, accounting for about 54% of the total useless soil evaporation.

4.5 Concluding remarks

This chapter focuses on the variations of energy/water budget and WUE in different agro-ecosystems. For the typical irrigated wheat and maize cropland in the NCP with careful management, water and energy balances are mainly influenced by the planting system and its phenological characteristics and irrigation under similar meteorological conditions. On the multi-year scale, the average annual ΔW can be assumed as zero verified across wheat-maize cropland, pear orchard, and cotton field ecosystems. Among the three major agro-ecosystems, ET, drainage, and WD of the vegetable field were the highest and those of the cotton field were the lowest. However, WUE of the maize season was the highest of these crops, considering both of the ET and the yield. Therefore, although crops like orchards and vegetables exhibit higher economic value, the blind expansion of orchards and vegetable planting just for economic aims should be stopped. Due to the largest planting area of wheat-maize cropland, water-saving management of wheat and maize field ecosystem is still the key point. The balance of water input and output would be achieved by transitioning from annual double cropping system into quadruple harvest of three years.

Furthermore, we quantified the ET structure of the major agro-ecosystems that the average annual T/ET ranged between 60% and 70%, indicating that about 210–280 mm of the total water was lost as unproductive soil evaporation. In order to better understand crop water utilization in the most important wheat and maize cropland, we used isotope and micro-meteorological methods to determine the water uptake depth. The main water uptake depth of wheat and maize crops is above 40 cm, and the evaporation depth is above 20–30 cm. These results suggest considerable potential for decreasing evaporation by subsurface irrigation or reducing the irrigation wetting depth to 40 cm from the traditional 100 cm. It is proposed that the optimal drip tapes spacing of SSDI was 100 cm and the optimal burial depth was 30 cm. Compared with FI, the yield of SSDI treatment does not decrease significantly and 1,500 m^3/ha of water can be saved, accounting for 54% of the ineffective soil evaporation. Based on the total cultivated land area of 1.7 million hectares of Hebei Province, 2.55 billion m^3 water can be saved for the whole province, which has great significance for the control of groundwater overexploitation. Besides the SSDI, lots of new irrigation methods have large potential in water saving, for example, the tube-sprinkler irrigation, pillow irrigation, and small size border irrigation. By optimizing irrigation scheduling, the water productivity can be obviously improved.

References

Craig H, Gordon L. 1965. Isotopic oceanography: deuterium and oxygen 18 variations in the ocean and the marine atmosphere. In: Tongiori, E. (Ed.), *Proceedings of the Conference on Stable Isotopes in Oceanographic Studies and Paleotemperatures.* Laboratory of Geology and Nuclear Science, Pisa, Italy, pp. 9–130.

Keeling C D. 1961. The concentration and isotopic abundances of carbon dioxide in rural and marine air. *Geochimica et Cosmochimica Acta*, 24(3–4): 277–298.

Lin G H, Sternberg L D L. 1994. Utilization of surface water by red mangrove (*Rhizophora mangle* L.): an isotopic study. *Bulletin of Marine Science*, 54(1): 94–102.

Min L, Shen Y, Pei H, Wang P. 2018. Water movement and solute transport in deep vadose zone under four irrigated agricultural land-use types in the North China Plain. *Journal of Hydrology*, 559: 510–522.

Phillips D L, Gregg J W. 2003. Source partitioning using stable isotopes: coping with too many sources. *Ecosystems Ecology*, 136(2): 261–269.

Phillips D L, Newsome S D, Gregg J W. 2005. Combining sources in stable isotope mixing models: alternative methods. *Oecologia*, 144(4): 520–527.

Philip J R. 1966. Plant water relations: some physical aspects. *Annual Review of Plant Physiology*, 17: 245–268.

Sun H, Liu C, Zhang Y, Zhang X. 2004. Study on soil evaporation by using micro-lysimeter. *Journal of Hydraulic Engineering*, 8: 1–6 (in Chinese).

Umair M. 2019. Quantification and reduction of soil evaporation loss in an irrigated cropland: a combined pathway of crop modelling and precision irrigation? PHD thesis, University of Chinese Academy of Sciences, Beijing.

Umair M, Shen Y, Qi Y, Zhang Y, Ahmad A, Pei H, Liu M. 2017. Evaluation of the CropSyst model during wheat-maize rotations on the North China Plain for identifying soil evaporation losses. *Frontiers in Plant Science*, 8: 1667.

Wang P, Li X, Wang L, Wu X, Hu X, Fan Y, Tong Y. 2018. Divergent evapotranspiration partition dynamics between shrubs and grasses in a shrub-encroached steppe ecosystem. *New Phytologist*, 219(4): 1325–1337.

Wang P, Song X, Han D, Zhang Y, Liu X. 2010. A study of root water uptake of crops indicated by hydrogen and oxygen stable isotopes: a case in Shanxi Province, China. *Agricultural Water Management*, 97(3): 475–482.

Wei Z, Lee X, Wen X, Xiao W. 2018. Evapotranspiration partitioning for three agro-ecosystems with contrasting moisture conditions: a comparison of an isotope method and a two-source model calculation. *Agricultural and Forest Meteorology*, 252: 296–310.

Wen X, Yang B, Sun X, Lee X. 2016. Evapotranspiration partitioning through in-situ oxygen isotope measurements in an oasis cropland. *Agricultural and Forest Meteorology*, 230: 89–96.

Xiao W, Wei Z, Wen X. 2018. Evapotranspiration partitioning at the ecosystem scale using the stable isotope method: a review. *Agricultural and Forest Meteorology*, 263: 346–361.

Yakir D, Sternberg L D L. 2000. The use of stable isotopes to study ecosystem gas exchange. *Oecologia*, 123(3): 297–311.

Yepez E A, Williams D G, Scott R L, Lin G H. 2003. Partitioning overstory and understory evapotranspiration in a semiarid savanna woodland from the isotopic composition of water vapor. *Agricultural and Forest Meteorology*, 119: 53–68.

Zhang X, Pei D, Chen S. 2004. Root growth and soil water utilization of winter wheat in the North China Plain. *Hydrological Processes*, 18(12): 2275–2287.

Zhang Y, Shen Y, Sun H, Gates J B. 2011. Evapotranspiration and its partitioning in an irrigated winter wheat field: a combined isotopic and micrometeorologic approach. *Journal of Hydrology*, 408(3–4): 203–211.

Zhang Y, Shen Y, Wang J, Qi Y. 2022. Estimation of evaporation of different cover types using a stable isotope method: Pan, bare soil, and crop fields in the North China Plain. *Journal of Hydrology*, 613: 128414.

Zuo Q, Sun, Y, Yang P. 1998. Study on water uptake by the winter wheat's roots with the application of Microlysimeter. *Journal of Hydraulic Engineering*, 6: 69–76 (in Chinese).

Chapter 5

Improving water productivity by integrating bio-, agro- and engineering measures

Xiying Zhang

5.1 Introduction: importance of WP

Fresh water crisis due to water shortages and water pollution is a serious problem around the world. Grain production, a sector heavily dependent on irrigation, is facing unprecedented challenges. Further increase in grain production to meet the requirements of a growing population would put further pressure on water resources in the world in future. Producing more food with less water by increasing water use efficiency (WUE) or water productivity (WP) is an important measure to solve the water crisis in agriculture (De Fraiture and Wichelns, 2010).

WP at yield level is defined as grain produced per unit water consumption, that is, evapotranspiration (ET), during the crop growth season (WP = yield/ET) (Fernández et al., 2020). The average WP of the three main grain crops in China was 0.85 kg/m^3 for rice, 1.01 kg/m^3 for wheat and 1.51 kg/m^3 for maize (Li and Peng, 2009). A higher WP results in either the same production from less water consumption or a higher production from the same water use. Zwart and Bastiaanssen (2004) reviewed 84 literature sources with results of experiments within the last 25 years, and they found that the average WP of wheat, rice and maize was 1.09, 1.09 and 1.80 kg/m^3, respectively, around the world. Zwart and Bastiaanssen (2004) also reported there was a large range of WP for the same crop (wheat, 0.6–1.7 kg/m^3; rice, 0.6–1.6 kg/m^3; and maize, 1.1–2.7 kg/m^3). The gap between the low WP and the maximum WP was quite large, and thus offers tremendous opportunities for maintaining or increasing agricultural production using less water, in particular by improving WP for grain production while reducing irrigation water use in water shortage regions.

The improvements in WP depend on the relative changes in grain production and crop water consumption. Grain yield (GY) of crop highly depends on the above-ground biomass and harvest index (HI). Thus, the WP at biomass level refers to the relation between biomass at maturity growth stage and ET. HI is defined using the following equation: HI=GY/Biomass. Hence, WP at GY level is equal to WP at biomass level multiplied by HI. Therefore, the increase in biomass and HI or the decrease of ET would increase WP at the GY level. HI is affected both by genetic background and agronomic management practices. HI had increased due to the revolution of plant, facilitated by shifting the early tall varieties to short varieties (Hütsch and Schubert, 2018). In the USA the maize yield increased by plant breeding and improved management (Duvick, 2005). Nutrients and water input are needed in biomass production; therefore, the management of nutrients and irrigation affect the WP of a crop. Similarly, crop ET is composed of crop transpiration (T) and soil evaporation (E). Soil E doesn't participate in the photosynthesis process, and should be reduced to save water and increase WP. Optimizing irrigation scheduling based on the water sensitivity of the crops during different growing stages can help to maximize

DOI: 10.1201/9781003221005-7

WP and stabilize yields under limited water supply (Zhang et al., 2008, 2013a). The suitable combination of irrigation timing, amount per application and irrigation frequency under different irrigation methods plays an important role in irrigation management.

Improving WP for grain production in the North China Plain (NCP) is an urgent priority. NCP is one of the most important grain production areas in China. The rich soil and climate are favourable for growing winter wheat and summer maize as a double cropping system. The mean annual rainfall is about 450–600 mm. About 70% of the rainfall occurs from July to September, the growing season for maize. Rainfall during the winter wheat growing season, which is from October to May of the following year, ranges from about 60 to 180 mm. Supplemental irrigation is required to support wheat production, as the consumptive water use by winter wheat is about 430–500 mm (Zhang et al., 2011). Farmers in this region generally irrigate winter wheat three or four times each season. They also irrigate maize one to three times per season. Farmers have pumped significant amounts of groundwater in recent years, contributing to a sharp decline in the groundwater table. A recent report entitled "Water shortages loom as Northern China's aquifers are sucked dry" reported the serious water shortage problem in the NCP (Li, 2010). An increase in WP can be achieved by improved agronomic practices, irrigation technologies, breeding and their integrated management. This chapter summarizes the effects of biological, agronomic and engineering measures on WP improvement for winter wheat and maize, the two major crops in the NCP. The results were based on field experimental results carried at Luancheng Eco-Agro-Experimental Station of the Chinese Academy of Sciences (37°53′N, 114°40′E; 50 m above sea level) (simplified as Luancheng Station), which is located in the middle of the NCP.

5.2 WP improvement for winter wheat and summer maize for the past four decades

A field experiment conducted at Luancheng Station from 1980 to 2019 indicated a significant improvement in WP for winter wheat and maize (Table 5.1). The increase in the yield of winter wheat was 53.1% and maize 59.7% during the past four decades. During the same period, WP of the two crops increased by 30% for winter wheat and 60% for maize. The significant increase in WP means that with the significant increase in yield, water use did not simultaneously increase. The improved WP guaranteed a continuous increase in crop yield with limited water input, which is quite important for regions with water shortage problems. The improved WP was the combined effects of cultivar renewing, reducing soil evaporation by mulch and improved soil fertility (Zhang et al., 2005, 2017, 2021).

Table 5.1 Average GY, seasonal water consumption (ET), WP for winter wheat and maize during the past four decades (1980–2019) for a same field at Luancheng Station in the NCP

Decades	Winter wheat			Summer maize		
	ET (mm)	Yield (kg/ha)	WP (kg/m^3)	ET (mm)	Yield (kg/ha)	WP (kg/m^3)
1980–1989	398.8	4695.9	1.18	374.60	5169.2	1.39
1990–1999	423.7	5631.3	1.32	381.88	7179.9	1.87
1999–2009	455.9	6639.5	1.46	390.16	7760.5	1.98
2010–2019	455.2	7187.9	1.60	373.14	8254.8	2.23

The improvement in the WP for the past four decades was associated with the renewing cultivars and the increase in soil nutrient contents. The yield improvement by cultivar renewal for winter wheat and maize was significantly related to the increase in HI, resulting in the increase in WP. The average HI was 0.38–0.40 in the 1980s, and it was 0.45–0.48 currently, indicating there was about 15%–20% increase in HI for winter wheat. The average HI was 0.38 for summer maize in the 1980s, and now it is around 0.57. The increase in HI for maize from the 1980s to the present was about 48%, which is greater than that for winter wheat, and is the reason that the WP increase during the past four decades was greater for summer maize than for winter wheat.

Biomass production requires large amounts of water and nutrients. The increase in HI would allow a more stable yield, as well as improved water and nutrient use efficiency. HI has a high heritability and is less affected by environmental factors. Maize usually has HI ranging from 0.25 to 0.58. The values for modern varieties usually fall within the range of 0.4–0.6. The HI improvement contributed 64% of the WP increase for maize and 30% for winter wheat (Zhang et al., 2013a, 2021). Breeding efforts had positive effects on biomass production and its partitioning. Without apparent increase in water use, the yield improvement driven by biomass or HI had increased WP. Results from Curin et al. (2020) demonstrated that breeding efforts had unintended positive effects on radiation and WUE, which are welcome on a global scale, considering that the predicted trends of water shortage will require increased WP rather than increased water use by crops.

Currently, the HI of the two crops maintains a relative stable value. Further HI improvements would require breakthroughs in new varieties and management skills in the NCP. Hütsch and Schubert (2017) suggested using different phytohormones to either enhance the activity of H+-ATPase for better kernel setting or reduce the shoot growth for higher HI of maize. Genetic or agronomic improvements to HI are becoming more important for higher resource use efficiency under resource-limited environments.

The changes in management practices, especially the straw return from both crops for the annual double cropping system in the NCP, significantly increased the soil organic matter (OM). Crop straw return has been widely recognized as an effective practice to manage carbon sequestration in agricultural ecosystems. The soil fertility increase at the experimental site was also related to the increase in chemical fertilizer use, especially urea. Together with the improved soil OM, soil N contents also significantly increased. At the experimental site of this study the soil OM and N contents were significantly increased during the past four decades. The average values of available N, P and K contents in the tillage layer increased by 2.6, 1.5 and 2.1 mg/kg/year, respectively. OM in the tillage layer increased to 23.9 g/kg in 2022 from 15.0 g/kg in 2002 due to long-term straw return practices and the increase in chemical fertilizer input (Li et al., 2023). Many studies have found that the increase in soil fertility and chemical fertilizer input also contributed greatly to the yield improvement of crops.

There are many benefits of increasing soil OM on crop production, which include the increased nutrient holding capacity of soil, improved water infiltration, decreased soil evaporation, preventing soil compaction and encouraging root development (e.g. Dexter, 2004). Studies have found that the linkage between nutrient mass flow and transpiration, particularly that of NO_3^-, partially regulates plant water flux (Cramer et al., 2009). Gloser et al. (2007) found that a rapid decrease in root hydraulic resistance occurred in the presence of increased nitrate availability. This trait might help to enhance a plant's ability to compete for nitrate and water in the soil. Thus, under water deficit condition, the increase in soil N content might increase the drought resistance of the crop. There are also studies showing that the increase in N supply improved the intrinsic water use efficiency (WUE_i) of plants (Köhler et al., 2012). Thus, the improved yield and WUE_i were also influenced by the increase in soil N availability for the two crops.

Another aspect in WP improvement for the two crops during the past four decades was related to the reduction in soil evaporation. The increased biomass production for the two crops would result in a large canopy size, and the increased canopy coverage would reduce soil evaporation. Liu et al. (2002) indicated that soil evaporation for summer maize in the NCP was ~30% of the total seasonal ET. Zhang et al. (2005) indicated that using the straw from winter wheat to mulch summer maize could reduce soil evaporation up to 40 mm during a growing season. Therefore, the increased canopy coverage and the use of straw mulch all contributed to the improved WP in the NCP for the two crops.

5.3 The contribution of breeding in WP improvements

Improvements in crop yield by breeding have been well reported. In China, for instance, an average wheat yield improvement of 52% has occurred due to genetic improvements since the 1980s (Zhang et al., 2013b). Wang et al. (2011) showed that genetic gain from 1964 to 2001 in summer maize was 69 kg/ha/year in China. Breeding to improve the morpho-physiological and/or agronomic features has greatly contributed to GY increases. Yield improvement through crop breeding has shown a huge impact on increasing WP.

5.3.1 Cultivar renewing on WP

Introducing new cultivars (cultivar renewal) played an essential role in yield and WP improvement in the NCP. Studies had been carried out to understand the genetic gains in yield and WP and their associated physiological and agronomic traits for winter wheat (Zhang et al., 2009, 2013b). Field experiment was conducted including ten cultivars of winter wheat released between 1970 and 2000 to examine the effects of breeding on yield and WP improvement. The ten cultivars were local popular cultivars and widely cultivated during the time when they were released. Same growing conditions were applied to those cultivars, with three irrigation levels arranged, that is, with one, two and three times irrigation application for two growing seasons in 2006–2009. Results showed that the average yield was improved by 24%–26%, while ET slightly increased under less irrigation condition, and maintained similar value under adequate water supply conditions for the cultivars released in different years, which resulted in a significant increase in WP from 15% under water deficit condition up to 25% under adequate water supply condition (Table 5.2). The results indicated that with the breeding of new cultivars, crop water use didn't increase, but GY was significantly improved. The increase in crop production without concomitant increase in ET by breeding new cultivars significantly improved the WP in crop production.

The genetic gains in GY for cultivars released in different years were associated with the increase in biomass, HI and kernel numbers per spike (Figure 5.1). Correlations were apparent between WP, HI and root : shoot ratio. Flowering date advanced by about 4 days over the 30 years of crop breeding period. Due to the limited grain-fill stage duration for winter wheat in the NCP, the advance in flowering time would prolong the grain-fill duration. Carbohydrates produced during the grain-fill stage provide the most parts of grains for cereal crops (Dordas, 2012). Post-anthesis dry matter production contributes more than 60% of the final GY for wheat

Table 5.2 Average seasonal ET, GY, WP of winter wheat cultivars released from 1970 to 2000 grown under the same condition with three irrigation treatments

Cultivar release period	One irrigation			Two irrigations			Three irrigations		
	ET (mm)	GY (kg/ha)	WP (kg/m³)	ET (mm)	GY (kg/ha)	WP (kg/m³)	ET (mm)	GY (kg/ha)	WP (kg/m³)
1970s	295.9	3926.9	1.32	380.1	4389.7	1.16	452.6	4710.9	1.04
1980s	316.3	4367.4	1.38	378.0	4787.4	1.27	433.1	5095.8	1.18
1990s	324.5	4959.5	1.53	389.5	5454.9	1.40	451.1	5874.4	1.30
Changes	9.7%	26.3%	15.1%	2.5%	24.3%	20.8%	−0.03%	24.7%	25.0%

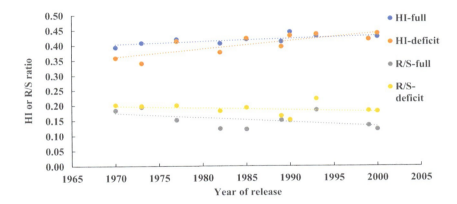

Figure 5.1 The changes in HI, root/shoot ratio (R/S) for winter wheat cultivars released in different years grown under two irrigation conditions (deficit irrigation and full irrigation).

(Masoni et al., 2007; Yang and Zhang, 2006). Therefore, longer grain-fill stage would benefit biomass production as well as the translocation of the dry matter to grains, which would benefit a higher HI. The increase in HI would greatly benefit the WP at GY level. Under the same growing condition, cultivars with a higher HI would use water more efficiently.

With less carbohydrate located to root growth, WP also improves with a smaller root : shoot ratio (Figure 5.1). When the cultivars released in different years were grown under the same conditions, the yield of recently released cultivars was greater than that of the earlier released cultivars under the three irrigation treatments (Table 5.2). However, the ET under more frequently irrigated treatments changed less from earlier to recently released cultivars. Under deficit water condition (one irrigation treatment), the ET of recently released cultivars was greater than that of the earlier released cultivars. The greater ET under the water deficit condition for new cultivars indicates that the new cultivars could extract more soil water than the old cultivars. The greater ability to extract soil moisture depends on the root system, either on the size of root system or on the ability of water uptake, or both (Zhang et al., 2004). Since root size is not a decisive factor in crop water use, smaller roots would be more economical. Total root length decreased from earlier to recent released cultivars and was significantly correlated with plant height. The breeding of winter wheat that reduced plant height not only increased HI, but also reduced root size, resulting in a smaller root : shoot ratio, which favoured a higher WP. The reduction in total root length from earlier to recent released cultivars mainly occurred in the top soil profile. The results indicate that total root length is not a factor that determines soil water use; rather, the distribution of root length density (RLD) through the soil profile plays a more important role in soil water utilization.

The results from the ten cultivars of winter wheat released in different years showed that the crop breeding programme has not only increased GY, but also produces grain more economically. The breeding programme both reduced plant height and increased the HI. For the underground root system, the reduction in plant height significantly reduced root size, and as a consequence, the root : shoot ratio. Although the root size was reduced, the water extracting ability by roots was not affected. The reduction in root size mainly occurred in the top soil profile, where RLD is greater, and it was not a limiting factor for crop water uptake (Li et al., 2022). Thus, with a high HI and a lower root : shoot ratio, more dry matter was allocated to above-ground biomass, resulting in improved grain production and WP, with the renewal in cultivars.

5.3.2 Selecting better cultivars to improve yield and WUE

Historically, cultivar renewal contributed greatly to yield and WP improvement. With the intensifying in water shortage, and under the climate change background, new crop varieties needed to be developed to have physiological or agronomic advantages that are suited to future growing conditions. Among the newly released cultivars, their performance in yield and WP often differs significantly. Researchers have shown that under water-limited conditions, the GY variation in cultivars ranges from 20.0% to 71.8% (Lopes and Reynolds, 2010; Lu et al., 2020; Zhang et al., 2005, 2009, 2010). The morpho-physiological features among cultivars are closely related to the mechanisms and strategies of drought resistance under dry conditions. Characteristics related to drought resistance, such as crop canopy temperature (CT), photosynthetic rate, chlorophyll, carbon isotope discrimination ($\Delta^{13}C$), biomass production and allocation efficiency, ash content and root distribution, have been widely studied (Deery et al., 2016; Hlavá ová et al., 2018). However, due to the genotype, environment and genotype × environment interactions, selectable traits might produce opposite results under different environments (Lu et al., 2020).

A four-season experiment from 2016 to 2020 on ten recently released winter wheat cultivars which are widely grown in the same region were conducted to assess the yield and WP among the cultivars and traits that are related to higher yield and WP. The ten cultivars were treated with adequate nutrient application, with three levels of irrigation treatments, that is, no irrigation during the growing season, one irrigation application at the jointing stage and two irrigations at jointing and flowering stages. Under the three irrigation treatments, seasonal ET reached about 60%, 70% and 80% of the seasonal potential ET, respectively, indicating the cultivars grown from serious water deficit to light water deficit conditions. Table 5.3 shows the average yield of the ten cultivars, the highest and the lowest yield among the ten cultivars, and the difference between the highest and the lowest yield for the four seasons under the three irrigation treatments. The results indicated there was large seasonal variation in crop production, and yield variations among the cultivars were greater than 100% in some seasons. The results indicated that there was large difference in cultivar performance in grain production, and the difference among the cultivars were also affected by the seasonal weather conditions. Selecting a better cultivar could offset some negative effects of water deficit and weather factors on crop production.

Table 5.3 Average yield, highest and lowest yields, yield variation among the ten recently released cultivars of winter wheat grown under three irrigation treatments from 2016 to 2020 at Luancheng Station

Irrigation levels	Yield items (kg/ha)	2016–2017	2017–2018	2018–2019	2019–2020
No irrigation	Average	5969.1	4647.1	2490.9	3104.0
	Highest	6724.4	4921.4	2882.1	4161.5
	Lowest	5302.8	4265.9	1852.3	1513.7
	Variations	26.8%	15.4%	55.6%	174.9%
One irrigation	Average	6767.3	5234.2	4876.8	5089.6
	Highest	7731.0	5631.2	5713.6	5838.3
	Lowest	5887.2	4951.9	4172.3	2810.3
	Variations	31.3%	13.7%	36.9%	107.7%
Two irrigations	Average	7830.8	5342.2	7045.1	6502.0
	Highest	8522.3	5546.0	7608.8	7530.3
	Lowest	7000.1	5111.3	6324.8	3833.0
	Variations	21.7%	8.5%	20.3%	96.5%

Using less water to produce more grain is very important in regions with water shortages. A significant positive linear relationship was found between WP and the GY among the ten winter wheat cultivars, as shown in Figure 5.2. The process of breeding to select high-yielding cultivars also improves WP. Therefore, selecting a high-yielding cultivar not only improves grain production but also increases WP.

Selecting cultivars with beneficial traits to improve yield and WP would be very important for maintaining grain production under water shortage conditions. For current winter wheat cultivars, the agronomic and physiological characteristics being related to the higher GY and higher WP included higher post-anthesis dry matter accumulation and higher final above-ground

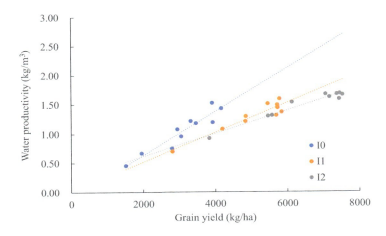

Figure 5.2 Correlation of GY with WP for different winter wheat cultivars grown with three irrigation levels (I0: no irrigation; I1: one irrigation applied at jointing stage; I2: additional irrigation added during flowering after I1).

biomass. Therefore, the agronomic and physiological characteristics after anthesis were related to the final yield of different cultivars. In a high-yielding environment, cultivars with relatively low CT tended to produce higher yields than those with high CT during grain-fill stage. A significant relationship was also found between GY and chlorophyll content at the late grain-filling stage. The high leaf chlorophyll content during that time would sustain the photosystem capacity of the crop to promote the post-anthesis dry matter accumulation. A negative linear relationship between the anthesis date and the HI among the current cultivars indicated that the earlier flowering cultivars had the advantage of prolonging the duration of grain filling, and the longer grain-fill duration was beneficial for improving the HI and WP. The above-mentioned traits could be used for selecting cultivars of winter wheat with better performance in yield and WP (Fang et al., 2017; Lu et al., 2020).

Root-related drought-adaptive characteristics, such as root biomass, root distribution and root/shoot ratio, are mainly affected by genetic background and year of release (Fang et al., 2017; Zhang et al., 2009). Richards (1991) and Richards et al. (2007) confirmed that early root vigour in crops may be related to a deeper root system in the field. A uniform and deeper root architecture reduces water use in the vegetative stage and increases access to water from deeper soil layers during reproductive growth stage (Palta et al., 2011). The RLD may help a plant maximize its use of resources and contribute to the production of biomass and a higher yield in dryland environments where crops depend on deeper soil water during grain-filling stage (Saradadevi et al., 2015). Thus, maximal soil moisture is captured for transpiration and minimal water is lost by soil evaporation, thereby increasing the WUE. Cultivars with uniform and deep root architectural features may be better in dryland environments.

GY of summer maize in the NCP increased from 5,054 kg/ha in the 1980s to 7,874 kg/ha in the 2000s due to increased HI and above-ground biomass, and prolonged grain-fill duration (Wang et al., 2011; Zhang et al., 2011). However, yield of summer maize has great inter-annual variation and it increased marginally in recent years due to the fluctuating climate. Sun et al. (2016) showed that variability of sunshine hours and the diurnal temperature range during grain-fill stage were associated with the seasonal yield variation. The reduced solar radiation would limit yield potential of irrigated summer maize in the NCP (Liu et al., 2010). Lower temperature and sunshine hours at vegetative stage of summer maize would result in small source size and less kernel numbers. Wang et al. (2014) indicated that adoption of new crop varieties was an effective measure to stabilize and increase maize yield under unfavourable conditions.

Selecting a good performing cultivar would be an effective way to reduce the negative effects of weather conditions on GY of maize to achieve high WP, due to the positive correlation of GY and WP among cultivars. Liu et al. (2010) showed that autonomous adoption of new maize cultivars in the NCP was able to stabilize the length of pre-flowering period and extend the length of the grain-fill period. Results from maize variety test at Luancheng Station indicated that a better performing cultivar could reduce the maize seasonal yield variation up to 32%. Agronomic evaluations of different cultivars are important to determine the traits of the better performing cultivars. Generally, crops with canopy structure to intercept more solar radiation may benefit from dry matter accumulation (Stewart et al., 2003). Leaf angle has significant influences on light distribution in the canopy. Cultivars with smaller leaf angle of upper leaves had higher GY (Fang et al., 2019; Liu et al., 2023).

Significant positive relationships between GY and above-ground biomass were found among the maize cultivar assessed at Luancheng Station. Cultivars with higher biomass accumulation benefited from the improvement in GY. The higher transfer efficiency of dry matter to kernel grains was also important to obtain a higher yield. HI was positively related to GY among different cultivars, but biomass contributed more to GY than to that of HI. Due to the shortened growth duration for summer maize following winter wheat in the northern part of NCP, cultivars with earlier tasselling date had an advantage to prolong the grain-fill duration and increase grain weight. Newer commercial maize cultivars tended to have longer duration of the grain fill. Selecting cultivars with earlier tasselling date and fast grain-fill speed would decrease the seasonal yield variation and maintain a high and stable yield of summer maize in the NCP.

5.4 Optimizing irrigation management under limited water supply

It is necessary to produce the maximum yield per unit area by using available water efficiently because irrigation water is rapidly diminishing around the world. At present and more so in the future, irrigated agriculture will take place under water scarcity. Irrigation management will shift from emphasizing production per unit area towards maximizing the production per unit of water consumed, the WP (Chaves and Oliveira, 2004; Fereres and Soriano, 2007; Geerts and Raes, 2009; Kang et al., 2017). It is essential to develop the most suitable irrigation schedule and adopt effective irrigation methods to produce the maximum plant yield. For winter wheat in the NCP, the seasonal rainfall is around 120 mm, which is far below the 450 mm water requirements. Irrigation is quite important for high yields of this crop. Optimizing the irrigation schedule and using suitable irrigation methods for winter wheat in the NCP would be quite important to save irrigation water in this region.

5.4.1 The relation of seasonal ET with WP and yield

The biological or primary productivity is normally evaluated in terms of biomass (as dry matter), and the water consumed and not recoverable in this production process is normally assessed in terms of ET, the sum of transpiration by the crop (T) and evaporation from the soil (E). The linearity between crop biomass (and often final yield) and water use has been observed since the early 1900s, and several hundreds of linear relationships can be found in the literature along with different approaches to link both variables. There are also studies showing that ET may not result in proportionate yield increases. Recently, a lot of researches showed that more than optimized water supply resulted in the decrease of both GY and WP. A long-term irrigation study at Luancheng Station showed that dry matter production, GY and WP were not linearly related to ET, but were best described in a quadratic curve (Figure 5.3). The maximum total dry matter production at maturity was achieved at 94% of the seasonal full ET and the highest GY was produced at 84% of the seasonal full ET (Zhang et al., 2008, 2013b, 2017). A positive relationship was found between HI and dry matter remobilization during grain filling. Moderate water deficit accelerated the remobilization and transfer of dry matter from vegetative tissues to the grain, resulting in higher GY and WP. The more rapid transfer of dry matter to the grain is of particular importance under the climatic conditions of the NCP where grain filling only lasts for one month because of the weather conditions.

Generally higher WP occurred with relatively lower ET. The maximum WP was achieved with a water supply of about half of the potential ET, but yield was much reduced under this ET. To optimize yield while achieving a higher WP, a reasonable water supply could be applied to save irrigation water.

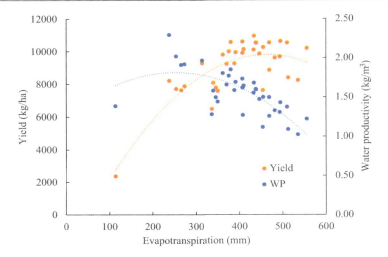

Figure 5.3 The correlation of seasonal ET with yield and WP for winter wheat.

5.4.2 Deficit irrigation scheduling for winter wheat

To cope with scarce water supplies, deficit irrigation, defined as the application of water below full crop water requirements (ET), is an important tool to achieve the goal of reducing irrigation water use (Fereres and Soriano, 2007). Extensive studies have demonstrated that post-anthesis water deficits result in early senescence and more mobilization of pre-anthesis stored assimilates to grains in cereals. Delayed senescence retards mobilization and can lead to reduced grain weight and much nonstructural carbohydrate left in the straw when growing seasons are terminated (Yang, 2015). It was concluded that optimization of irrigation inputs at the farm level would maximize biomass production as well as increase the HI and WP.

Under a limited water supply, it is very important to decide how to schedule supplemental irrigation to minimize yield penalty. In the NCP, before winter wheat sowing, the rainy season usually replenishes some of the soil water, and the water stored in the soil profile plays an important role in supplying water to the winter wheat. The seasonal rainfall, stored soil water and supplemental irrigation comprise the water sources for winter wheat in this region. Results from nine-season study, comprising four treatments: rain-fed, single irrigation applied at sowing to obtain a good level of soil moisture at the start of crop growth (I1s), single irrigation applied during recovery to jointing (I1r) and full irrigation supplied with three irrigations (control, I3), showed that GY significantly correlated with rainfall before heading and with ET after heading ($P < 0.01$) under rain-fed conditions. The average contribution of soil water stored before sowing to seasonal ET was 90, 103 and 145 mm for rain-fed, I1s and I1r, respectively, during the six seasons. The single irrigation applied around the time of recovery to jointing (I1r) produced more GY than the irrigation applied at the earlier stage of the winter wheat development (Zhang et al., 2013a). This practice resulted in a greater biomass accumulation before heading and more efficient soil water use during grain fill, two factors that were significantly related to GY under a limited water supply.

A smaller RLD, which restricted the utilization of deep soil water by the crop, was one of the reasons for the lower yield with rain-fed and I1s treatments compared with the I1r treatment in dry seasons. The results also showed that the limited irrigation applied from recovery to jointing stage (Treatment I1r) significantly promoted vegetative growth and more efficient soil water use during the reproductive (post-heading) stage, resulting in a 21.6% yield increase compared with that of the I1s treatment (Figure 5.4). Although the average yield of the I1r treatment was 14% lower than that of the full irrigation treatment, seasonal irrigation was reduced by 120–140 mm. With smaller penalties in yield and a larger reduction in applied irrigation, I1r could be considered a feasible deficit irrigation practice that could be used in the NCP for conservation of groundwater resources (Zhang et al., 2013a).

Results from this study also showed that soil water depletion contributed approximately 40%–50% of the total ET under rain-fed and limited water supply conditions. The seasonal rainfall cannot be controlled, so improving soil water use by developing the crop root system would contribute greatly to yield improvement of winter wheat under a limited water supply (Zhang et al., 2004, 2009, 2013a). The assumption that a crop would develop a deep and large root system under drought was not always true. The results from this study showed that during the dry seasons, the rain-fed treatment extracted less soil water because the smaller RLD in deep layers of the soil limited the crop's ability to fully extract the available soil water (Figure 5.5). The results from this study showed that for all treatments, the RLD declined rapidly with an increase in soil depth. The RLD in the top soil layer was relatively greater. Zhang et al. (2009) found that reducing root growth redundancy in the top soil layer and enhancing the root's ability to deplete more soil water are clearly adaptive features for wheat in water-limited conditions. For the situation in the NCP, enhancing root growth in deep layers of the soil is important for efficient soil water use under limited water supply conditions. The results from this study showed that relatively vigorous growth during the vegetative growth period promoted root growth that enhanced soil water utilization in the later growth period of winter wheat.

Comparing the I1r and I1s treatments, the timing of the limited supplemental irrigation significantly affected the yield of winter wheat during dry seasons, due to the timing of limited

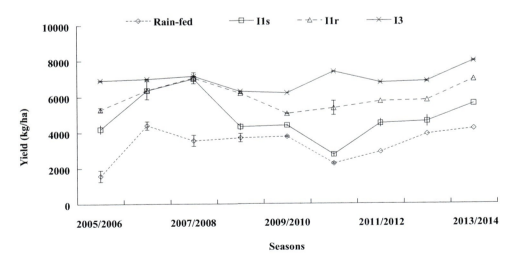

Figure 5.4 The average yield of winter wheat under four irrigation treatments (rain-fed, one irrigation applied earlier, I1s; and later, I1r; three irrigations, I3) for six seasons from 2005 to 2014 (the bars represent the standard deviation). (Modified based on Zhang et al., 2013a.)

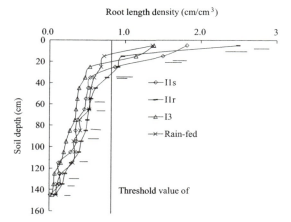

Figure 5.5 The RLD distribution at anthesis under four irrigation treatments (rain-fed, one irrigation applied earlier, I1s; and later, I1r; three irrigations, I3) (the threshold value of the RLD indicated that when the RLD was smaller than 0.8 cm/cm³, the root was the restricting factor for full utilization of the soil water by the crop) (Zhang et al., 2013a).

water supply that affected its total availability to crop water use and its availability at different growing stages of crops, which influenced the biomass production, GY and HI. Grain production could be improved under terminal drought if the water use during the vegetative growing stages was reduced, and soil water availability during grain-fill stage was increased (Ma et al., 2008; Yan et al., 2020). The supplemental irrigation from recovery to jointing for I1r maintained higher soil moisture before anthesis compared to that of I1s. Biomass vigour also promoted root growth that enhanced soil water use during grain filling compared to that of I1s. The final yield and WP from I1r was substantially improved over I1s, although the two treatments had similar irrigation application amounts.

In the NCP, full irrigation of winter wheat can guarantee a higher and more stable GY. However, the high seasonal water consumption resulted in a rapid decline of the groundwater table. Under this full irrigation practice, agricultural water use was not sustainable. The results from the above-discussed study showed that with some yield penalty under a limited water supply, ET could be significantly reduced to a level at which sustainable GY and agricultural water use could be achieved. Optimizing the limited water supply could minimize yield reduction. The summer season in the NCP is the rainy season, and the soil moisture along the root-zone profile can be replenished to some extent before sowing winter wheat. By regulating the timing of the application of a single irrigation, soil water use could be improved resulting in substantially improved crop production during the dry seasons. Sustainable irrigation water use could be achieved with smaller penalties in yield and a larger reduction in crop water use in the NCP.

5.4.3 Using suitable irrigation methods

The above results indicated that a single irrigation at jointing stage to winter wheat could reduce irrigation water use by 120–140 mm as compared with full irrigation treatment, but with yield penalties of 14% for winter wheat in the NCP. Further study should be required to find solutions to minimize the yield reduction under limited water supply. Optimizing the irrigation strategies is one of the major measures to efficiently utilize limited irrigation water (Rodrigues and Pereira, 2009). Some studies indicated that frequent application of water in small quantities can provide appropriate water for crop growth, increasing WUE (Enciso et al., 2003). However, traditional irrigation methods such as flooding irrigation may not easily control the irrigation amount per application (Sezen et al., 2006). Irrigation methods that can control the irrigation frequency and amount per application might be required to improve the irrigation WUE for field-grown crops, such as winter wheat (Bian et al., 2016). Micro-irrigation can precisely control the irrigation application and is popular in orchards, vegetable growing and wide-row planted crops.

Drip irrigation is one of the major micro-irrigation methods, allowing accurate application of irrigation in small amounts directly to the root zone (Bhunia et al., 2015). Increasing the frequencies by drip irrigation could maintain a high soil matric potential in the root rhizosphere, accelerating crop growth, decreasing deep percolation, reducing soil evaporation and saving irrigation water (Jensen et al., 2014; Qin et al., 2016). Other irrigation methods such as pillow irrigation (PI) and tube-sprinkler irrigation (SI) are also being used in growing crops. PI, which uses plastic tubes with holes on the tubes and the tubes are placed between two rows of the crop, can directly deliver water to the crop. It has the benefits of drip irrigation and plastic mulching. This method can also restrict weed growth and improve GY as well as WP under limited water supply (Gerçek et al., 2017). For the application of the micro-sprinkler irrigation to the densely planted crops, such as winter wheat, a modified micro-sprinkler irrigation is becoming popular in China now. A thin tube is placed on the soil surface with holes on it. The pressured water will come out from the holes acting as sprinkler. Micro-irrigations can save water compared with flood irrigation due to those methods decreasing the water loss from evaporation, deep percolation as well as during water delivery. With the increase in irrigation water shortage, water-saving irrigation methods would become more important in the future. However, their wide use in crops is still limited, due to the high investment and maintaining cost of the irrigation system and the lower economic returns of the crops (Paredes et al., 2014).

Three micro-irrigation methods, PI, SI and drip irrigation (DI), were compared with the local traditional basin irrigation (BI) method under limited water supply at Luancheng Station to evaluate the effects of irrigation amount/frequency and different irrigation methods on yield and WP of winter wheat. Under the same limited irrigation amount (90 mm/season), two irrigation applications (45 mm/application) from recovery to grain fill conducted using DI method significantly increased the yield and WP of winter wheat as compared with the BI method using one single application (90 mm/application) at jointing stage (Table 5.4) (Fang et al., 2018). The improved yield and WP of winter wheat under DI with the increased irrigation frequency were related to the increased water availability to crops during the grain-fill stage, resulting in improved seed weight. Increasing the seasonal irrigation amount to 160 mm, the increase in the application frequency by reducing the irrigation amount per application did not significantly affect the yield using either PI or SI methods. Under the same irrigation management, the DI outperformed the SI system in yield and WP partly due to the possible reduction in soil evaporation. The results showed that limited water supply irrigation method that can reduce soil evaporation and increase the irrigation water availability to crops has the advantage of improving both yield and WP. However, with adequate irrigation water supply, the irrigation method did not affect the yield, but may improve the WP by reducing soil evaporation. Due to the high cost in installation of the three micro-irrigation systems, the net income was reduced by 30% as compared with the

Table 5.4 Crop water use (ET), soil water depletion (SWD), yield and WP under different irrigation methods for three seasons of winter wheat[a]

Seasons	Treatments	Irrigation (mm)	Irrigation numbers	SWD (mm)	Rainfall (mm)	ET (mm)	Yield (kg/ha)	WP (kg/m^3)
2012–2013	Rain-fed	0	0	156.6	122.2	278.8	4630.5	1.66
2013–2014	Rain-fed	0	0	215.2	47.9	263.1	5361.0	2.04
2014–2015	Rain-fed	0	0	150.3	77.1	227.4	3447.0	1.52
2012–2013	Basin	90	1	137.4a	122.2	349.6a	5772.0b	1.65b
	Drip	90	2	119.3b	122.2	331.5b	6462.0a	1.95a
2013–2014	Basin	90	1	244.9a	47.9	382.8a	7038.0b	1.83b
	Drip	90	2	223.3b	47.9	361.2b	7947.5a	2.20a
2012–2013	Basin	160	2	109.9a	122.2	392.1a	6817.5a	1.74b
	Pillow	160	3	65.0b	122.2	347.2b	6840.0a	1.97a
	Tube-sprinkler	160	3	99.6a	122.2	381.8a	6837.0a	1.79b
2013–2014	Basin	160	2	223.9a	47.9	431.8a	8218.5a	1.90a
	Pillow	160	3	203.4b	47.9	411.3b	7988.6a	1.94a
	Tube-sprinkler	160	3	222.4a	47.9	430.3a	8103.0a	1.88a
2014–2015	Basin	160	2	236.7a	77.1	473.8a	7381.5a	1.56d
	Basin	90	1	228.8a	77.1	395.9b	6217.5d	1.57d
	Drip	90	2	195.3b	77.1	362.4c	6937.8b	1.91a
	Pillow	90	2	232.7a	77.1	399.8b	6898.3b	1.73b
	Tube-sprinkler	90	2	238.5a	77.1	405.6b	6614.5c	1.63c

[a] Under the same column and in the same season, values followed with the same letter were not significant at $P = 0.05$.

BI method. The economic water productivity ratio (EWPR) was only three to four for the three micro-irrigation systems, much less than the BI method, which had an average value of 16. Currently, the BI method is more economic for growing winter wheat in the NCP.

The improvement in the performance of winter wheat was related to the increased top soil water contents during the grain-fill stage which significantly increased the seed weight under DI. Some studies showed that frequent irrigation probably resulted in the increase in soil evaporation (Sebastian et al., 2016). The results from this study showed that soil evaporation was increased under SI as compared with the BI-2 with the increased irrigation frequency. However, due to the partly wetted soil surface under DI, soil evaporation was slightly reduced under DI as compared with that of BI-1. Although SI increased soil evaporation, the benefit from the increased irrigation frequency improved the soil water status during the grain fill and then the final crop production under limited irrigation supply condition.

The root system of winter wheat can reach 2 m in NCP (Zhang et al., 2004). However, the smaller RLD in deep soil profile restricted extraction of soil water depletion. Meanwhile, the growth of deep root system would also consume photosynthetic products (Kashiwagi et al. 2015). Generally, most roots of cereal crops are concentrated in the top 40 cm soil layer where water and nutrients are abundant, thus reducing the cost in root growth to absorb nutrients and water. It has been reported that when plants grow in productive environments, they have a faster return of investment (de la Riva et al., 2021). In contrast, plants that survive in a resource-limiting environment tend to have a slower return of investment (Roumet et al., 2016). The coupling of water and nutrients in the top soil layer by increasing the irrigation frequency using micro-irrigation methods reduced the cost in soil water and nutrient uptake and enhanced the WP in biomass production.

Adopting optimized irrigation scheduling and irrigation methods had the advantages to increase WP and save irrigation water. But for farmers the economic return is one of the most important issues to be considered. Table 5.5 shows that the net income from using the micro-irrigation systems was much less than that using the BI system, due to the high cost in installation and maintenance of the micro-irrigation systems. The performance of different systems should be also assessed in its economic outcomes (Paredes et al., 2014). The total cost of DI was higher than surface irrigation, even with a higher benefit in water saving (Rodrigues et al., 2013). Although DI and PI produced higher WP and GY as compared with the BI method under the same seasonal irrigation amount, the higher investment in the equipment offset the benefit.

Table 5.5 The economic outcome and the EWPR of different irrigation treatments averagely for the three seasons from 2012 to 2015

Treatments[a]	Yield (kg/ha)	Gross income (Yuan/ha)	Irrigation cost (Yuan/ha)	Other costs (Yuan/ha)	Total cost (Yuan/ha)	Net income (Yuan/ha)	EWPR
Rain-fed	4479.5	9854.9	---	4950.0	4950.0	4904.9	---
BI-1	6342.5	13953.5	898.0	4950.0	5848.0	8105.5	15.54
BI-2	7472.5	16439.5	1052.0	4950.0	6002.0	10437.5	15.63
DI	7115.8	15654.7	4422.0	4950.0	9372.0	6282.7	3.54
PI	7242.3	15933.1	4798.4	4950.0	9748.4	6184.7	3.32
SI	7184.8	15806.6	3703.4	4950.0	8653.4	7153.3	4.26

[a] BI-1, basin irrigation with one application; BI-2, basin irrigation with two applications; DI, drip irrigation; PI, pillow irrigation; SI, tube-sprinkler irrigation.

Consequently, the net income of micro-irrigation methods was lower than that of BI. The lower net income would restrict the application of those water-saving irrigation methods in practice. Surface adequate irrigation (irrigation amount at 160 mm) produced the highest net income among the different treatments, indicating the benefits in reasonable increase in the irrigation water use using the low-investment irrigation method. The results from this study also showed that the net income was greater even using the expensive irrigation methods than that of the rain-fed treatments. The results indicated the importance of supplemental irrigation to winter wheat in the NCP. In places with serious water shortage problem, the adoption of the micro-irrigation system to improve the crop production might be feasible.

EWPR was a useful indicator because it combined GY value and the total costs in irrigation. The cost of the applied water, transportation equipment, the GY response to water and price of the agricultural product decide the economic benefits of irrigation systems (Rodrigues et al., 2013). Chandran and Surendran (2016) indicated that if the farm area became larger, the initial investment cost will be lower due to the reduction of the components of irrigation system, such as the filter, pump set, head unit and tank. Rodrigues et al. (2013) indicated that the feasibility of DI depending on the commodity prices and the irrigation system was far from economic viability due to the low commodity prices. Higher commodity prices could lead to positive incomes. Therefore, increasing the grain price of winter wheat and water price was a useful method that could increase the net income using micro-irrigation systems, thus increasing the willingness of the farmers' adoption in water-saving methods. Since the reduction in irrigation water use and at the same time maintaining the grain production capability are the two important issues in the NCP, local government might need to increase the subsidy to farmers to use the water-saving irrigation methods for the benefit of sustainable agricultural development in the region.

Increasing irrigation frequency by micro-irrigation systems could improve GY and WP under limited water supply. Under adequate water supply condition, more frequent irrigation did not affect the GY. The three micro-irrigation methods mentioned above all improved yield and WP under limited water supply. However, due to the high investment in the irrigation system, the net income was significantly reduced as comparing with the BI method. Currently, BI methods were more economic than that of the other three micro-irrigation systems in the NCP. But with water scarcity intensifying, the increase in farm size and the possible subsidy from government, adoption of those water-saving irrigation methods might be feasible in the future.

5.5 Reducing soil evaporation

WP is defined as yield/ET, ET is composed of crop transpiration (T) and soil evaporation (E). Reducing E will decrease the ET and increase the WP. The double cropping system of winter wheat and maize in the NCP requires over 850 mm water annually, in which 20%–30% is from E that is generally thought non-effective for crop production (Liu et al., 2002). For winter wheat, average total seasonal ET was 453 mm under full water supply, in which E took up 29.8% of the seasonal ET. Evaporation varies appreciably at different stages. Before winter dormancy (October and November), E/ET is 44%; during winter dormancy (December, January and February) it rises to 68.9%. This ratio then drops to 31% in recovery, and remains at 27% during the jointing to the grain-filling stages. During the period from heading to grain filling, evaporation reaches a minimum of 17%, and then slightly rises to 18% at maturity (Liu et al., 2002).

For summer maize, average total seasonal ET was 422 mm under full water supply, in which E was about 30.3% of the ET. At seedling stage, the E/ET was around 65%, and with the quick growth of the crop, this ratio rapidly decreased to around 22% till the maturity stage.

At maturity, E/ET slightly increased to 25% (Liu et al., 2002). In the NCP, the two crops per year rotation system of winter wheat and summer maize, about 263 mm water consumption, is from soil evaporation under full water supply condition. Therefore, reducing soil evaporation could perform a very important role in increasing the WP and reducing the irrigation water use.

5.5.1 Factors affecting E/ET

When the soil water supply can meet the transpiration needs of crops, E/ET is mainly affected by leaf area index (LAI) and surface soil moisture. With the increase in canopy size, the ratio of E/ET decreases markedly. A correlation of E/ET with LAI was developed to describe the correlation of the two factors under full water supply (Liu et al., 2002). For winter wheat: $E/ET = 1/(1 + 0.5(LAI) + 0.18(LAI^2))$ and for maize $E/ET = 1.12 \exp(-0.34 \, LAI)$. The equation shows that when moisture is not the limiting factor, the E/ET is mainly controlled by LAI.

Topsoil moisture greatly influences the E/ET ratio. The water used for field ET comes from the whole root zone, while the soil evaporation is controlled by topsoil moisture. Though the moisture in the root zone can meet the requirements for transpiration, the topsoil moisture has already affected soil evaporation. Generally, E/ET fell with the decrease in topsoil moisture. The relation could be expressed by the following equation when LAI being 3: $E/ET = 1.16 + 0.53(Ln\theta_v)$, where θ_v is the volumetric moisture of topsoil. The change of E/ET (at LAI = 3) accompanying the decrease in topsoil moisture can be divided into several phases. When the topsoil moisture is at the field capacity, E/ET remains steady with a value at about 50%, then it drops slowly with the loss of topsoil moisture. Whenever the topsoil moisture falls to the capillary water broken point, E/ET drops sharply to about 10% with the loss of the topsoil moisture until the topsoil moisture is lowered to a certain degree. From then on, E/ET would remain at a very low but constant level. The results suggest that an important step to reduce soil evaporation is to keep topsoil dry without affecting the crop transpiration (Zhang et al., 2005; Fang et al., 2021).

5.5.2 Reducing soil evaporation by mulch

Straw mulching has been widely used to conserve soil moisture. Studies have shown that straw mulching can prevent the soil surface from receiving direct sunlight, reduce the absorption of radiation and the movement of soil water to reduce evaporation from soil surface (Li et al., 2018; Akhtar et al., 2019). The double cropping system of winter wheat and maize produces a large quantity of straw every year that is a good source for straw mulching. Experiments on straw mulching were carried out at Luancheng Station for over ten years to investigate the effects of straw mulching on soil evaporation, soil temperature and crop production as well as WP.

Micro-lysimeters were used to measure daily soil evaporation with and without straw mulching at Luancheng Station. Results showed that the average soil evaporation rate for mulched treatment was smaller than that of non-mulched treatment, especially during the earlier growing stages of two crops when LAI was low. Figure 5.6 shows the daily soil E with mulch and without mulch for maize with the changes in LAI during a growing season. Soil E was significantly reduced with mulch when LAI was smaller for maize. When LAI was greater, the reduction in E was reduced, but during the whole growing period, E under mulch was always lower than that without mulch. The results over a 12-year period of measuring for maize showed that straw mulch reduced soil evaporation by 40–50 mm and improved WP by 7%–10%. For winter wheat under straw mulching, soil evaporation was reduced by 40%, averaged over five seasons. Straw

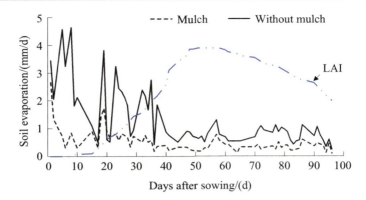

Figure 5.6 The changes in daily soil evaporation with the LAI during the maize growing season with and without mulch for a typical season at Luancheng Station (Zhang et al., 2005).

mulching reduced soil evaporation by about 80–100 mm annually for the two crops. With the use of combines to harvest wheat, straw mulch could be readily applied to maize.

Straw mulch on the soil surface also affects soil temperature which in turn influences crop growth, especially in winter crops. For maize, the summer crop, the mulch did not affect the crop development and yield. In contrast, with winter wheat the presence of straw on the soil surface reduced soil temperature during the period of critical development. From February to early April winter wheat is during the recovery and jointing stages. During this time the average daily soil temperature (0–10 cm) was 0.42°C and 0.65°C lower under light and heavily mulched conditions, respectively, compared to a non-mulched control (Chen et al., 2007; Rashid et al., 2019). On average, over five seasons the lower soil temperature under mulch in spring delayed the development of winter wheat up to five days and reduced the final GY by 5% and 7% for light and heavily mulched treatments, respectively, as compared with the control. After April with the increase in leaf area the effect of mulch on soil temperature gradually disappeared. Although soil evaporation was reduced under mulch, yield of winter wheat was reduced. The overall WP was not changed under straw mulching. It seems that under these experimental conditions the negative impact of delayed maturity was greater than any benefit of reduced evaporation from the soil surface for winter wheat, which is grown during the cool season. While for summer maize, which is grown during the hot summer season, straw mulch in conserving soil moisture benefits both grain production and WP improvements.

Higher plant transpiration in mulching treatment improved maize growth (increased LAI and biomass) and water utilization under deficit water supply (Fang et al., 2021). Soil mulching converts more water from unproductive soil evaporation to plant transpiration. In addition, Thidar et al. (2020) showed that straw mulching improved root growth on the top 30 cm soil layers, which also benefit water uptake and increase GY under rain-fed conditions. Straw mulching not only increased the soil moisture content but also improved soil fertility and enzyme activities (Akhtar et al., 2018), Furthermore, the straw materials are inexpensive, convenient and friendly to the environment (Kader et al., 2017). Application of straw mulching will be an effective measure to increase crop yield under limited water supply during warm seasons.

5.6 Further increase WP in future

With intensifying water shortage and the increase in food demand with population growth in China, further increase in grain production will become more important. Further improvement in WP and efficient utilization of the limited water resources is important to mitigate the groundwater overdraft problem in the NCP.

The mechanisms that underlie the responses of crops to water deficit involve many processes such as intercellular CO_2, oxidative stress, sugar signalling, membrane stability and root chemical signals. In water-limited environments, photosynthetic carbon gain and loss of water by transpiration are in a permanent trade-off as both are oppositely regulated by stomatal conductance. Large unregulated fluxes of water are not essential to plant functioning and water can be saved by manipulating stomatal aperture. Future work will be required to continue focusing on understanding the factors that regulate the trade-off between carbon assimilation and water loss, and those that drive the partitioning of assimilates between reproductive and non-reproductive structures in relation to soil water availability. Rhizosphere manipulation, especially partial root-zone drying, root hydraulic resistance in response to nitrate supply and other methods to alter root to shoot signalling to regulate crop growth and water loss will also be important factors to relieve the negative effects of water stress on crop yield.

Deficit irrigation strategies are likely increasingly to be adopted around the world due to the water shortage problem. The responses of crops to water stress are affected by crop type, cultivar type and phenological stage as well as the crop growing conditions. With the frequent change in cultivars and growing environments of grain crops, the opportunities for applying deficit irrigation practices and related strategies need to be continuously investigated and developed to fit different practical situations.

Continual breeding of new cultivars has significantly improved grain production and WP in China over the past four decades. Considering the vast population, limited arable land per farmer, the relative low education background and weak economic situation in rural area, the introduction of new cultivars is much easier than any other technologies in farming. Every year there are many new cultivars developed and sold in markets. However, farmers don't know which cultivar is more fitting to their conditions and what kinds of field management measures are required for a special cultivar. Studies are needed to focus on the traits of different cultivars that are beneficial to yield and WP improvement. Corresponding field management practices in irrigation, fertilizing and cultivation should be developed to bring out the yield and WP potentials of the new major cultivars of winter wheat and maize in NCP.

References

Akhtar K, Wang W, Khan A, Ren G, Afridi M Z, Feng Y, Yang G. 2019. Wheat straw mulching offset soil moisture deficient for improving physiological and growth performance of summer sown soybean. *Agricultural Water Management*, 211: 16–25.

Akhtar K, Wang W, Ren G, Khan A, Feng Y, Yang G. 2018. Changes in soil enzymes, soil properties, and maize crop productivity under wheat straw mulching in Guanzhong, China. *Soil & Tillage Research*, 182: 94–102.

Bhunia S, Verma I M, Arif M, Gochar R, Sharma N C. 2015. Effect of crop geometry, drip irrigation and bio-regulator on growth, yield and water use efficiency of wheat (*Triticum aestivum* L.). *International Journal of Agricultural Sciences*, 11(1): 45–49.

Bian C, Ma C, Liu X, Gao C, Liu Q, Yan Z, Ren Y, Li Q. 2016. Responses of winter wheat yield and water use efficiency to irrigation frequency and planting pattern. *PloS One*, 11(5): e0154673.

Chandran K M, Surendran U. 2016. Study on factors influencing the adoption of drip irrigation by farmers in humid tropical Kerala, India. *International Journal of Plant Production*, 10(3): 347–364.

Chaves M M, Oliveira M M. 2004. Mechanisms underlying plant resilience to water deficits: prospects for water-saving agriculture. *Journal of Experimental Botany*, 55(407): 2365–2384.

Chen S, Zhang X, Pei D, Sun H, Chen S. 2007. Effects of straw mulching on soil temperature, evaporation and yield of winter wheat: field experiments on the North China Plain. *Annals of Applied Biology*, 150(3): 261–268.

Cramer M D, Hawkins H, Verboom G A. 2009. The importance of functional regulation of plant water flux. *Oecologia*, 161: 15–24.

Curin F, Severini A D, Gonzalez F G, Otegui M E. 2020. Water and radiation use efficiencies in maize: breeding effects on single-cross Argentine hybrids released between 1980 and 2012. *Field Crops Research*, 246: 107683.

De Fraiture C, Wichelns D. 2010. Satisfying future water demands for agriculture. *Agricultural Water Management*, 97(4): 502–511.

De la Riva E G, Prieto I, Maranon T, Perez-Ramos I M, Olmo M, Villar R. 2021. Root economics spectrum and construction costs in Mediterranean woody plants: the role of symbiotic associations and the environment. *Journal of Ecology*, 109(4): 1873–1885.

Deery D M, Rebetzke G J, Jimenez-Berni J A, James R A, Condon A G, Bovill W D, Hutchinson P, Scarrow J, Davy R, Furbank R T. 2016. Methodology for high-throughput field phenotyping of CT using airborne thermography. *Frontiers in Plant Science*, 7: 1808.

Dexter A R. 2004. Soil physical quality: part I. Theory, effects of soil texture, density, and organic matter, and effects on root growth. *Geoderma*, 120: 201–214.

Dordas C. 2012. Variation in dry matter and nitrogen accumulation and remobilization in barley as affected by fertilization, cultivar, and source-sink relations. *European Journal of Agronomy*, 37(1): 31–42.

Duvick D N. 2005. The contribution of breeding to yield advances in maize (*Zea mays* L.). *Advances in Agronomy*, 86: 83–145.

Enciso J M, Unruh B L, Colaizzi P D, Multer W L. 2003. Cotton response to subsurface drip irrigation frequency under deficit irrigation. *Applied Engineering in Agriculture*, 19(5): 555–558.

Fang Q, Wang Y, Uwimpaye F, Yan Z, Liu X, Shao L. 2021. Pre-sowing soil water conditions and water conservation measures affecting the yield and water productivity of summer maize. *Agricultural Water Management*, 245: 106628.

Fang Q, Zhang X, Chen S, Shao L, Sun H. 2017. Selecting traits to increase winter wheat yield under climate change in the North China Plain. *Field Crops Research*, 207: 30–41.

Fang Q, Zhang X, Shao L, Chen S, Sun H. 2018. Assessing the performance of different irrigation systems on winter wheat under limited water supply. *Agricultural Water Management*, 196: 133–143.

Fang Q, Zhang X, Chen S, Shao L, Sun H, Yan Z. 2019. Selecting traits to reduce seasonal yield variation of summer maize in the North China Plain. *Agronomy Journal*, 111(1): 343–353.

Fereres E, Soriano M A. 2007. Deficit irrigation for reducing agricultural water use. *Journal of Experimental Botany*, 58(2): 147–159.

Fernández J E, Alcon F, Diaz-Espejo A, Hernandez-Santana V, Cuevas M V. 2020. Water use indicators and economic analysis for on-farm irrigation decision: a case study of a super high density olive tree orchard. *Agricultural Water Management*, 237: 106074.

Geerts S, Raes D. 2009. Deficit irrigation as an on-farm strategy to maximize crop water productivity in dry areas. *Agricultural Water Management*, 96(9): 1275–1284.

Gerçek S, Demirkaya M, Işik D. 2017. Water pillow irrigation versus drip irrigation with regard to growth and yield of tomato grown under greenhouse conditions in a semi-arid region. *Agricultural Water Management*, 180: 172–177.

Gloser V, Zwieniecki M A, Orians C M, Holbrook N M. 2007. Dynamic changes in root hydraulic properties in response to nitrate availability. *Journal of Experimental Botany*, 58: 2409–2415.

Hlaváčová M, Klem K, Rapantova B, Novotna K, Urban O, Hlavinka P, Smutna P, Horakova V, Skarpa E, …, Trnka M. 2018. Interactive effects of high temperature and drought stress during stem elongation, anthesis and early grain filling on the yield formation and photosynthesis of winter wheat. *Field Crop Research*, 221: 182–195.

Hütsch B W, Schubert S. 2017. Harvest index of maize (*Zea mays L.*): are there possibilities for improvement? *Advances in Agronomy*, 146: 37–82.

Hütsch B W, Schubert S. 2018. Maize harvest index and water use efficiency can be improved by inhibition of gibberellin biosynthesis. *Journal of Agronomy and Crop Science*, 204(2): 209–218.

Jensen C R, Orum J E, Pedersen S M, Andersen M N, Plauborg F, Liu F, Jacobsen S E. 2014. A short overview of measures for securing water resources for irrigated crop production. *Journal of Agronomy and Crop Science*, 200(5): 333–343.

Kader M A, Senge M, Mojid M A, Ito K. 2017. Recent advances in mulching materials and methods for modifying soil environment. *Soil & Till Research*, 168: 155–166.

Kang S, Hao X, Du T, Tong L, Su X, Lu H, Li X, Huo Z, Li S, Ding R. 2017. Improving agricultural water productivity to ensure food security in China under changing environment: from research to practice. *Agricultural Water Management*, 179: 5–17.

Kashiwagi J, Krishnamurthy L, Purushothaman R, Upadhyaya H D, Gaur P M, Gowda C L L, Ito O, Varsheny R K. 2015. Scope for improvement of yield under drought through the root traits in chickpea (*Cicer arietinum L.*). *Field Crop Research*, 170: 47–54.

Köhler I H, Macdonald A, Schnyder H. 2012. Nutrient supply enhanced the increase in intrinsic water-use efficiency of a temperate seminatural grassland in the last century. *Global Change Biology*, 18: 3367–3376.

Li B, Peng S. 2009. *Reports on agricultural water use in China from 1998-2007*. Beijing: China Agriculture Publishing House (in Chinese).

Li H, Li L, Liu N, Chen S, Shao L, Sekiya N, Zhang X. 2022. Root efficiency and water use regulation relating to rooting depth of winter wheat. *Agricultural Water Management*, 269: 107710.

Li H, Shao L, Liu X, Sun H, Chen S, Zhang X. 2023. What matters more, biomass accumulation or allocation, in yield and water productivity improvement for winter wheat during the past two decades? *European Journal Agronomy*, 149: 126910.

Li J. 2010. Water shortages loom as Northern China's aquifers are sucked dry. *Science*, 328(5985): 1462–1463.

Li S, Li Y, Lin H, Feng H, Dyck M. 2018. Effects of different mulching technologies on evapotranspiration and summer maize growth. *Agricultural Water Management*, 201: 309–318.

Liu C, Zhang X, Zhang Y. 2002. Determination of daily evaporation and evapotranspiration of winter wheat and maize by large-scale weighing lysimeter and micro-lysimeter. *Agricultural and Forest Meteorology*, 111: 109–120.

Liu N, Li L, Li H, Liu Z, Lu Y, Shao L. 2023. Selecting maize cultivars to regulate canopy structure and light interception for high yield. *Agronomy Journal*, 115(2): 770–780.

Liu Y, Wang E, Yang X, Wang J. 2010. Contributions of climatic and crop varietal changes to crop production in the North China Plain, since 1980s. *Global Change Biology*, 16(8): 2287–2299.

Lopes M S, Reynolds P M. 2010. Partitioning of assimilates to deeper roots is associated with cooler canopies and increased yield under drought in wheat. *Functional Plant Biology*, 37: 147–156.

Lu Y, Yan Z, Li L, Gao C, Shao L. 2020. Selecting traits to improve the yield and water use efficiency of winter wheat under limited water supply. *Agricultural Water Management*, 242: 106410.

Ma S, Xu B, Li F, Liu W, Huang Z. 2008. Effects of root pruning on competitive ability and water use efficiency in winter wheat. *Field Crop Research*, 105: 56–63.

Masoni A, Ercoli L, Mariotti M, Arduini I. 2007. Post-anthesis accumulation and remobilization of dry matter, nitrogen and phosphorus in durum wheat as affected by soil type. *European Journal Agronomy*, 26(3): 179–186.

Palta J A, Chen X, Milroy S P, Rebetzke G J, Dreccer F, Watt M. 2011. Large root systems: are they useful in adapting wheat to dry environments? *Functional Plant Biology*, 38: 347–354.

Paredes P, Rodrigues G C, Alves I, Pereira L S. 2014. Partitioning evapotranspiration, yield prediction and economic returns of maize under various irrigation management strategies. *Agricultural Water Management*, 135: 27–39.

Qin S, Li S, Kang S, Du T, Tong L, Ding R. 2016. Can the drip irrigation under film mulch reduce crop evapotranspiration and save water under the sufficient irrigation condition? *Agricultural Water Management*, 177: 128–137.

Rashid M A, Zhang X, Andersen M N, Olesen J E. 2019. Can mulching of maize straw complement deficit irrigation to improve water use efficiency and productivity of winter wheat in North China Plain? *Agricultural Water Management*, 213: 1–11.

Richards R A. 1991. Crop improvement for temperate Australia: future opportunities. *Field Crop Research*, 26: 141–169.

Richards R A, Watt M, Rebetzke G J. 2007. Physiological traits and cereal germplasm for sustainable agricultural systems. *Euphytica*, 154: 409–425.

Rodrigues G C, Paredes P, Gonçalves J M, Alves I, Pereira L S. 2013. Comparing sprinkler and drip irrigation systems for full and deficit irrigated maize using multicriteria analysis and simulation modelling: ranking for water saving vs. farm economic returns. *Agricultural Water Management*, 126: 85–96.

Rodrigues G C, Pereira L S. 2009. Assessing economic impacts of deficit irrigation as related to water productivity and water costs. *Biosystems Engineering*, 103(4): 536–551.

Roumet C, Birouste M, Picon-Cochard C, Ghestem M, Osman N, Vrignon-Brenas S, Cao K, Stokes A. 2016. Root structure-function relationships in 74 species: evidence of a root economics spectrum related to carbon economy. *New Phytologist*, 210(3): 815–826.

Saradadevi R, Bramley H, Palta J A, Edwards E, Siddique K H M. 2015. Root biomass in the upper layer of the soil profile is related to the stomatal response of wheat as the soil dries. *Functional Plant Biology*, 43: 62–74.

Sebastian B, Lissarrague J R, Santesteban L G, Linares R, Junquer, P, Baeza P. 2016. Effect of irrigation frequency and water distribution pattern on leaf gas exchange of cv. 'syrah' grown on a clay soil at two levels of water availability. *Agricultural Water Management*, 177: 410–418.

Sezen S M, Yazar A, Eker S. 2006. Effect of drip irrigation regimes on yield and quality of field grown bell pepper. *Agricultural Water Management*, 81(1–2): 115–131.

Stewart D W, Costa C, Dwyer L M, Smith D L, Hamilton R L, Ma B L. 2003. Canopy structure, light interception, and photosynthesis in maize. *Agronomy Journal*, 95: 1465–1474.

Sun H, Zhang X, Wang E, Chen S, Shao L, Qin W. 2016. Assessing the contribution of weather and management to the annual yield variation of summer maize using APSIM in the North China Plain. *Field Crop Research*, 194: 94–102.

Thidar M, Gong D, Mei X, Gao L, Hao W, Gu F. 2020. Mulching improved soil water, root distribution and yield of maize in the Loess Plateau of Northwest China. *Agricultural Water Management*, 241: 106340.

Wang J, Wang E, Yin H, Feng L, Zhang J. 2014. Declining yield potential and shrinking yield gaps of maize in the North China Plain. *Agricultural Forest Meteorology*, 195: 89–101.

Wang T, Ma X, Li Y,Bai D, Liu C, Tan X, Shi Y, Song Y, …, Smith S. 2011. Changes in yield and yield components of single-cross maize hybrids released in China between 1964 and 2001. *Crop Science*, 51: 512–525.

Yan J, Bogie N A, Ghezzehei T A. 2020. Root uptake under mismatched distributions of water and nutrients in the root zone. *Biogeosciences*, 17(24): 6377–6392.

Yang J. 2015. Approaches to achieve high grain yield and high resource use efficiency in rice. *Frontiers of Agricultural Science and Engineering*, 2(2): 115–123.

Yang J, Zhang J. 2006. Grain filling of cereals under soil drying. *New Phytologist*, 169(2): 223–236.

Zhang X, Chen S, Liu M, Pei D, Sun H. 2005. Improved water use efficiency associated with cultivars and agronomic management in the North China Plain. *Agronomy Journal*, 97(3): 783–790.

Zhang X, Chen S, Sun H, Pei D, Wang Y. 2008. Dry matter, harvest index, grain yield and water use efficiency as affected by water supply in winter wheat. *Irrigation Science*, 27(1): 1–10.

Zhang X, Chen S, Sun H, Shao L, Wang Y. 2011. Changes in evapotranspiration over irrigated winter wheat and maize in North China Plain over three decades. *Agricultural Water Management*, 98(6): 1097–1104.

Zhang X, Chen S, Sun H, Wang Y, Shao L. 2010. Water use efficiency and associated traits in winter wheat cultivars in the North China Plain. *Agricultural Water Management*, 97(8): 1117–1125.

Zhang X, Chen S, Sun H, Wang Y, Shao L W. 2009. Root size, distribution and soil water depletion as affected by cultivars and environmental factors. *Field Crops Research*, 114(1): 75–83.

Zhang X, Pei D, Chen S. 2004. Root growth and soil water utilization of winter wheat in the North China Plain. *Hydrological Processes*, 18(12): 2275–2287.

Zhang X, Qin W, Chen S, Shao L, Sun H. 2017. Responses of yield and WUE of winter wheat to water stress during the past three decades: a case study in the North China Plain. *Agricultural Water Management*, 179: 47–54.

Zhang X, Uwimpaye F, Yan Z, Shao L, Chen S, Sun H, Liu X. 2021. Water productivity improvement in summer maize: a case study in the North China Plain from 1980 to 2019. *Agricultural Water Management*, 247: 106728.

Zhang X, Wang Y, Sun H, Chen S, Shao L. 2013a. Optimizing the yield of winter wheat by regulating water consumption during vegetative and reproductive stages under limited water supply. *Irrigation Science*, 31(5): 1103–1112.

Zhang X, Wang S, Sun H, Chen S, Shao L, Liu X. 2013b. Contribution of cultivar, fertilizer and weather to yield variation of winter wheat over three decades: a case study in the North China Plain. *European Journal Agronomy*, 50: 52–59.

Zwart S J, Bastiaanssen W G M. 2004. Review of measured crop water productivity values for irrigated wheat, rice, cotton and maize. *Agricultural Water Management*, 69(2): 115–133.

Part 3

Water adaptive agriculture
Reducing pumping intensity

Chapter 6

Agricultural land use change and impacts on groundwater

Yanjun Shen and Yucui Zhang

6.1 Introduction

Agricultural land use reflects the evolution of agricultural development driven by resources, technology, and socio-economic factors. Understanding patterns and changes in agricultural land use from a macro-perspective is critical to ensuring food security and sustainable land and water management. In the North China Plain (NCP), there are various types of agricultural land use for growing cereal crops, such as wheat, maize, millet, and cash crops such as cotton, peanuts, sesame, vegetables, and fruit trees (Zhang et al., 2019). The spatial distribution and change in time of agricultural land use are mainly determined by soil properties, water resource conditions, the profits of agricultural products, policies, etc. For example, wheat and maize double cropping system is distributed in the piedmont plain with adequate groundwater, cotton is mainly concentrated in the soil salinization areas, and vegetables are planted in the peripheral areas of cities (Zhang et al., 2020).

Over the past decades, the agricultural land use has experienced significant changes in the NCP. Before the 1980s, the agricultural cultivation pattern was dominated by cereal crops because the nation's primary concern at that stage was meeting the demand for ration food. Accompanied with growth of population and food demand, the scale of agricultural production has expanded significantly (Qi et al., 2022); cereal crops, especially wheat and maize, dominated the agricultural land use in the NCP. Agriculture gradually became intensive mostly benefited from the development of irrigation and use of chemicals since the 1980s. For the past 30 years, the irrigated cereal crops have still been the dominant agricultural land use, although the total planting area declined owing to the fast urbanization, which occupied and converted huge amount of arable land for urban use. It is worthy to note that land reclamation has also happened recently, especially in the saline soil region of eastern part of the NCP.

The planting area of cotton had significantly increased in the 1970s and 1980s, up to more than 10% of the total agricultural land, then decreased due to policies, price fluctuations, and cost increases. Because of higher economic profit, fruits and vegetables have expanded since 2000, consequently, water and fertilizer application rates increased too (Zhang et al., 2019).

The agricultural land use changes of past decades have witnessed a historical shift in cropping structure in the NCP, reflecting the socio-economic development and farmer's evolving objectives in agriculture. As a result of agricultural land use change, quantity and even quality of groundwater resources were profoundly impacted. Increases in area of water-consuming crops followed by heavier extraction of groundwater, and in most cases, more fertilizer applications, resulted in greater depletion and pollution risks of groundwater. These kinds of land use pattern changes in agricultural sector therefore could put compound impacts on aquifers in terms of quantity and quality, making the NCP a hotspot region globally.

Advances in recent decades of remote sensing technology give human a bird's-eye view to observe changes of land surface coverage and subsurface groundwater of aquifers. By remote sensing, some key agricultural information can be easily known, and further, agricultural, water-related policies had scientific decision-making support. Agricultural land use change is highly correlated with water use structure (Zhang et al., 2019), reflecting a region's macro-scale proportions of surface- and subsurface-water extraction amounts as well as rainfed and irrigated agricultural area. By monitoring these changes, water and agriculture sustainability can be effectively assessed. This chapter thus introduces agricultural land use change and its impact on agricultural water budget in the NCP, and discusses how to develop a water-suitable agriculture pattern.

6.2 Agricultural land use of the NCP in 2002 and 2012

Cropping system adjustment and agricultural production estimation are the key to ensure both food security and water saving. Monitoring the spatial pattern of crops through remote sensing technology can help to make clear the regional crop planting structure (crop types and planting areas), systems (multiple cropping or fallow), and patterns (continuous, rotation, or interplanting) (Tang et al., 2010). In order to reduce the inputs and maximize the profit, agricultural production strategies have changed dramatically in the NCP over the past decades.

6.2.1 Detection of agricultural land use change

Planting patterns and cropping systems are numerous in the NCP, and the overall crop distribution is fragmented and scattered. The distribution map of crops in NCP will be useful to clarify and optimize the planting pattern and agricultural disaster alleviation. Previous studies on the spatial and temporal pattern of crops in successive years usually used the visual interpretation method or were based on the mathematical analysis of statistical data (de Oliveira et al., 2017; Li et al., 2017; Liu et al., 2014). All these methods consumed lots of labour, material, and economic resources. Therefore, time- and labour-saving methods for crops planting area detection for the NCP in successive years are necessary.

According to field survey, statistical and phenological data collected from agricultural meteorological stations (Zhang et al., 2019), the main agricultural land use types include winter wheat-summer maize, single maize, cotton, and fruit orchard. It was much easier to classify crop types and extract the planting area according to their different phenological characteristics using remote sensing data (Figure 6.1b). We employed HANTS algorithms (Harmonic Analysis of Time Series, available from the Geospatial Data Service Centre (GDSC), Netherlands) (Roerink et al., 2000; White et al., 2009) to filter and treat the raw MODIS-NDVI data to reconstruct crops' NDVI time series for land use/crop classification. Classification and Regression Tree method (CART) (Breiman et al., 1984) was adopted for the detection of crop types over NCP (Pan, 2015) for years 2002 and 2012. Winter wheat-summer maize, single cropping maize, cotton, and forest/fruit trees were classified based on MODIS data, while the vegetables and rice are detected by using Landsat TM/ETM (Thematic Mapper/ Enhanced Thematic Mapper) data since these two types of land use are greatly fragmented and scatter distributed. Finally, we got the agricultural land use map as in Figure 6.2.

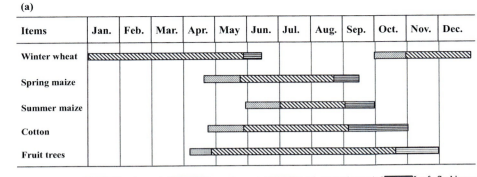

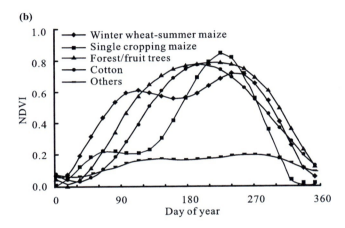

Figure 6.1 The main crops' growth periods (a) and NDVI time series curves of different crops (b) in the NCP.

The classification accuracy or reliability was assessed by comparing the classified areas of each type of land use with that of statistical yearbook at county level. The correlation coefficient between remote sensing classification results and agricultural statistics is higher than 0.77 for all the cropland use types (confidence interval: 95%, Figure 6.3). This method has high classification accuracy in crop extraction in the NCP, and can reflect the spatial and temporal changes of agricultural land use.

The distribution of different agricultural land use types of the NCP in 2002 and 2012 (Figure 6.2) shows winter wheat and summer maize are widely cultivated in south of Beijing and Tianjin, which are the main planting pattern in the NCP. The single maize is widely distributed in northern Hebei Province, especially in Langfang City. Cotton is mainly distributed in south of Hebei Province and northwest of Shandong Province, such as prefectures Xingtai, Hengshui, Dezhou, and Liaocheng. The fruit trees are concentrated in the areas of geologically depressed lowland places filled with sandy soil in historical floods in prefectures Shijiazhuang, Cangzhou, and Binzhou. The single crop maize is mainly cultivated near the big municipalities of Beijing and Tianjin, where the farmers intend to work in the cities for higher income and left the land

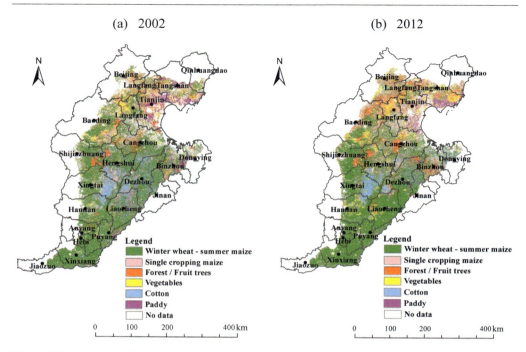

Figure 6.2 Agricultural land uses of the NCP in 2002 (a) and 2012 (b).

abandoned from wheat growing because of its low profit. On the other hand, farmers in these places are more likely to grow vegetables and fruits for relatively higher return, which could also be reflected from the agricultural land use types in these places.

6.2.2 Agricultural land use area changes

The total agricultural land use of the NCP was 8.93 million ha in 2002, and increased to 9.25 million ha in 2012, which was mainly caused by land reclamation of the formerly low-productive areas with soil salinization. An obvious increase in vegetables and forest/fruit trees planting area is also recognized (Table 6.1). As for the dominant winter wheat-summer maize double cropping system, the planting area accounted for about 55% of the total arable land area and a decrease of 34,200 ha observed from 2002 to 2012. Proportions of a total planting area of single spring maize, forest/fruit trees, and cotton are around 12% of the total arable land area. Planting area of cotton decreased for 221,400 ha, while planting areas of single spring maize and forest/fruit trees increased by 17.3% and 27.4% from 2002 to 2012, respectively. Vegetables accounted for only 6% of the total agricultural land area, observing considerable net increase from 447,600 ha in 2002 to 648,000 ha in 2012.

In order to clarify the differences of agricultural land use changes prefectures in NCP, the planting area changes from 2002 to 2012 were compared as shown in Table 6.2. The NCP was divided into four parts according to the administrative boundaries: Beijing and Tianjin Municipalities, Hebei, Shandong, and Henan Provinces.

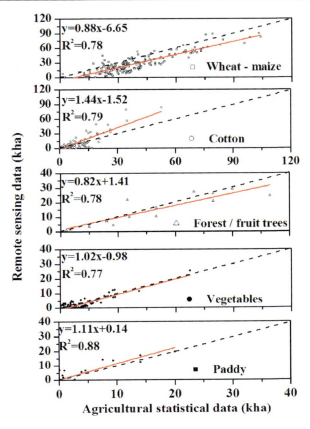

Figure 6.3 Comparison of classified land use area for different crops with statistical data (data points in the figures show area for years 2002 and 2012 at county level).

Grain crop planting areas of Beijing, Tianjin, and eastern Hebei Province (Qinhuangdao and Tangshan prefectures decreased obviously, while forest/fruit trees and vegetable planting areas increased. Because of the serious water shortage and groundwater limitation policies, winter wheat-summer maize planting areas in Langfang, Cangzhou, and Hengshui pectures decreased obviously and the average proportion reached 31.7%, and the total area reduced to 280,000 ha, during the decade of 2002–2012. However, the planting area of single spring maize increased about 160,000 ha, about 66.7%, compared with the previous planting area. The forest/fruit trees planting area also increased by about 40% in these three prefectures. Meanwhile, the planting area of vegetables doubled in Cangzhou and Hengshui prefectures. The cotton planted in Hengshui decreased by 22,740 ha. Such results do nothing to conserve groundwater. Grain crops planted in Baoding City also decreased while fruit trees increased, especially for several counties near Beijing City. The winter wheat-summer maize planting area in south of Hebei Province (including Shijiazhuang, Xingtai and Handan prefectures) had a few increase, because some farmers stopped planting other grains and switched to winter wheat and summer maize, due to more convenient sowing and harvest machines. The cotton in Shijiazhuang and Handan decreased by about 10,000 ha due to continued low prices, while vegetable planting area in Handan was doubled.

Table 6.1 Area and proportion of crops in the NCP

Year	Items	Winter wheat -summer maize	Cotton	Single spring maize	Forest/fruit trees	Vegetables	Paddy	Summation
2002	Area (10³ha)	5065.40	1087.37	1045.88	997.60	447.62	283.88	8927.75
	Proportion (%)	56.74	12.18	11.72	11.17	5.01	3.18	100.00
2012	Area (10³ha)	5031.21	865.90	1226.10	1271.17	648.02	216.51	9258.91
	Proportion (%)	54.34	9.35	13.24	13.73	7.00	2.34	100.00

Table 6.2 Agricultural land use change of municipalities or prefectures in the NCP from 2002 to 2012 (unit: 10^3 ha)

		Cotton	Single spring maize	Forest/fruit trees	Winter wheat	Vegetables	Paddy
	Beijing (BJ)	0.13	−21.09	69.96	−8.57	7.13	0.93
	Tianjin (TJ)	0.24	−5.09	−31.64	−19.91	48.92	−28.78
Hebei Province	Qinhuangdao (QHD)	0.72	13.78	14.63	−24.89	0.49	−2.36
	Tangshan (TS)	0.16	−31.07	35.38	−24.91	45.04	−25.12
	Langfang (LF)	8.78	56.61	33.18	−81.30	13.69	1.19
	Cangzhou (CZ)	9.61	63.08	34.28	−120.33	33.10	9.72
	Hengshui (HS)	−22.74	41.22	28.14	−78.82	23.27	0.56
	Baoding (BD)	8.06	−17.40	95.50	−38.88	−1.67	1.61
	Shijiazhuang (SJZ)	−10.62	−7.24	0.78	57.33	−1.72	0.01
	Xingtai (XT)	15.27	−29.04	−4.64	54.21	−0.51	0.00
	Handan (HD)	−9.36	10.71	−2.76	24.06	28.19	0.05
Shandong Province	Dongying (DY)	8.50	2.09	−7.61	10.46	−6.08	−10.58
	Binzhou (BZ)	−6.43	5.85	8.44	−23.92	−2.85	−4.35
	Jinan (JN)	10.26	2.99	0.73	−10.56	−0.07	1.80
	Dezhou (DZ)	−85.27	21.83	47.79	16.09	6.59	5.01
	Liaocheng (LC)	−139.00	12.08	−28.95	177.93	11.39	−7.18
Henan Province	Hebi (HB)	0.13	3.43	−0.28	7.53	−0.10	0.00
	Xinxiang (XX)	−7.24	17.90	0.65	12.56	−0.80	−4.07
	Jiaozuo (JZ)	0.29	5.49	1.32	−10.96	0.03	0.31
	Puyang (PY)	−0.52	21.79	−8.66	33.22	−0.81	−6.79
	Anyang (AY)	−2.34	12.38	−12.78	14.85	−2.11	0.01
	Total	−221.38	180.28	273.44	−34.82	201.15	−68.03

Water resources in Shandong Province are more abundant than in Hebei Province. Prefectures of Shandong Province in the NCP were near the Yellow River, which is more convenient to irrigate for farming. Therefore, the winter wheat cultivation did not decline significantly. The winter wheat cultivation in Liaocheng Prefecture increased by 177,930 ha during 2002–2012. At the same time, because of the cumbersome planting approach and low price, rice and cotton planting area were reduced in Liaocheng and Dezhou prefectures. More and more economic crops, such as watermelon, jujube, pear, and various vegetables, were planted in these prefectures. In Dezhou Prefecture, for example, cotton planting area decreased by 85,270 ha while the planting area of forest/fruit trees increased by 47,790 ha during the observation period. The land use change is not obvious in Dongying, Binzhou, and Jinan Prefectures, Shandong Province. Because of unused land reclamation in the flood-prone area of northern Henan Province, the planting area of winter wheat increased in this region. Similar situation happened in the coastal region of eastern Hebei Province.

In general, the changes of agricultural land use in Beijing, Tianjin Municipalities, and Henan Province were not significant, compared with the Shandong and Hebei Provinces. The land uses with increasing area in Beijing and Tianjin were crops with high economic benefit (fruit trees and vegetables, totally 94,368 ha). However, the planting area of grain crops (wheat and single maize) increased obviously in the Henan Province with a total of 118,181 ha. The significant changes in area of winter wheat and cotton cultivation were in Hebei and Shandong Provinces, which decreased more than 200,000 ha during 2002–2012. Meanwhile, the planting area of forest/fruit trees and winter wheat increased around 200,000 ha in these two provinces.

6.3 Effects of changes in winter wheat area on groundwater depletion

The winter wheat-summer maize double cropping system is the main planting pattern in the NCP, and it utilizes substantial groundwater resources for irrigation, resulting in severe environmental problems. An accurate estimation of crop water consumption and net irrigation water consumption is crucial to optimizing the management of agricultural water resources. A crop evapotranspiration (ET) estimation model was proposed, combining FAO Penman- Monteith method with remote sensing data (Wu et al., 2019).

The planting area of winter wheat has an indirect strong impact on water consumption. Therefore, the actual water consumption and net irrigation water consumption of winter wheat were estimated based on the proposed model (Wu et al., 2019). Meanwhile, the impact of changes of winter wheat planting area on groundwater depletion was analysed. These will provide valuable information for water management and the development of sustainable agriculture in the NCP.

6.3.1 Change of winter wheat area and ET

In order to estimate the water consumption of winter wheat, the planting area of winter wheat was retrieved using remote sensing data by the same method as in Section 6.2. At the same time, in order to better reflect the changes in winter wheat planting area, the research year was extended from 2012 to 2016. The NCP is divided into four parts: Northern Hebei Plain (N-HBP), Southern Hebei Plain (S-HBP), Northwestern Shandong Plain (NW-SDP), and Northern Henan Plain (N-HNP). Figure 6.4 illustrates the changes of the winter wheat planting area in the four sub-regions of the NCP during 2001–2016. At the sub-regional level, the growing area of winter wheat showed different trends from the period of 2001–2005 to 2012–2016. It had a fast

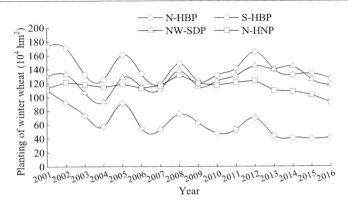

Figure 6.4 Change of winter wheat planting area in the four parts of the NCP during 2001–2016. N-HBP, North Hebei Plain; S-HBP, South Hebei Plain; N-HNP, North Henan Plain; NW-SDP, Northwest Shandong Plain.

decreasing trend and declined from 847×10^3 to 465×10^3 ha, a decrease of 45.1%, in the N-HBP; a mild decreasing trend in South Hebei Plain and North Henan Plain, with relative declinations of 8.2% and 8.5%, respectively; while the winter wheat planting area in the NW-SDP had an increasing trend and increased from 1.18×10^6 to 1.34×10^6 ha.

Evapotranspiration from agricultural land is generally affected by weather condition, crop type, soil moisture, etc. The model combining the FAO Penman-Monteith equation with the crop coefficient, and remote sensing data was developed to estimate actual crop ET. First, crop reference ET (ET_0) was calculated using the FAO Penman-Monteith equation. Crop water requirement (ETc) under standard conditions was calculated by multiplying ET_0 by the crop coefficient (Kc). The NCP is a typical water-stressed area, so the soil water condition is a crucial factor impacting actual ET of cropland. Soil moisture condition was taken into consideration when estimating actual crop ET in this study. The actual crop ET (ETa) was calculated using the following equation:

$$ETa = ET_0 \times Kc \times Ks \tag{6.1}$$

where ETa is the actual crop ET; ET_0 is the reference crop ET; Kc is the crop coefficient; and Ks is the soil moisture coefficient. We estimate Ks using remotely sensed NDVI data at each sub-region; by an assumption the Ks will be equal to 1 if the maximum NDVI in wheat growing season is larger than the value with 5% of exceedance, and nearly 0 if it is lower than the value with 95% of exceedance, then the Ks of each pixel could be estimated using its actual NDVI and the two threshold values through a simple linear conversion (Wu et al., 2019).

Based on the planting area of winter wheat, meteorological data, and remote sensing NDVI data, the amount of water consumption of winter wheat in the NCP was calculated (Figure 6.5a). The average values of water consumption of winter wheat during 2001–2016 were $2.05 \times 10^9 \, m^3$ in the N-HBP, $4.97 \times 10^9 \, m^3$ in the S-HBP, $4.85 \times 10^9 \, m^3$ in the Northwest Shandong Plain, and $4.65 \times 10^9 \, m^3$ in the N-HNP. Moreover, the water consumption of winter wheat in the N-HBP decreased from $2.97 \times 10^9 \, m^3$ during 2001–2005 to $1.22 \times 10^9 \, m^3$ during 2012–2016, a decrease of 56.0%. The water consumption of winter wheat in the S-HBP showed and decreased from $5.51 \times 10^9 \, m^3$ during 2001–2005 to $4.59 \times 10^9 \, m^3$ during 2012–2016. The reduction of winter wheat cultivation in the north and south of Hebei Plain is an important reason for decrease of water consumption by winter wheat.

6.3.2 Characteristics of net irrigation water consumption of winter wheat

Net irrigation water consumption of winter wheat was defined according to the water balance equation. The runoff can be ignored in water balance equation as it dropped to a virtually negligible level since the 1960s for the cropland in the NCP (Fan et al., 2010; Wu et al., 2011). The average soil water change for multiple years in the Luancheng Station was used as shown in Zhang et al. (2018). Therefore, the net consumed irrigation water (I_{net}) can be calculated as follows:

$$I - D = I_{net} = ET - P - \Delta SWC \tag{6.2}$$

where ET is the actual crop ET (mm), P is the precipitation (mm), I is the irrigation (mm), ΔSWC is the soil water change (mm), and D is the deep percolation (mm).

Figure 6.5b shows the spatial distribution of average precipitation in the winter wheat growing season from 2001 to 2016. The precipitation in the winter wheat growing season in the NCP ranged from 132 mm in the northwest to 215 mm in the southeast. In the piedmont plain of the Taihang Mountains, the precipitation was <180 mm and these regions need irrigation of approximately 150 mm. In most parts of the N-HBP and S-HBP, the precipitation in the winter wheat growing season was approximately <170 mm, and these regions need the irrigation of <100 mm. Compared with the Hebei Plain, the precipitation in the Northwest Shandong Plain and N-HNP were relatively higher and ranged from 170 to 215 mm. In the irrigation district of the Yellow River Basin in Henan and Shandong provinces, net irrigation water consumption was more than 150 mm.

Based on the planting area of winter wheat, the net irrigation water consumption in the four sub-regions of NCP from 2001 to 2016 was estimated (Figures 6.5c and 6.6). The average net irrigation water consumption from 2001 to 2016 was up to 800×10^6 m³ in the N-HNP, 1.99×10^9 m³ in the Southern Henan Plain, 1.78×10^9 m³ in NW-SDP, and 1.78×10^9 m³ in the N-HNP. The net irrigation water consumption in the N-HBP decreased from 1.10×10^9 m³

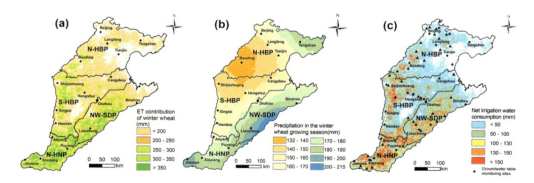

Figure 6.5 Spatial distribution of ET contributed from winter wheat field (a), precipitation (b), and net irrigation water consumption (c) in the winter wheat growing season in the NCP (2001–2016).

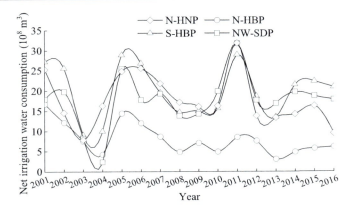

Figure 6.6 Change in net irrigation water consumption for the winter wheat growing season in the four sub-regions of the NCP during 2001–2016.

during 2001–2005 to $550 \times 10^6\,\text{m}^3$ during 2012–2016, mainly caused by decrease in wheat growing area. The net irrigation water consumption in the S-HBP decreased from $2.02 \times 10^9\,\text{m}^3$ during 2001–2005 to $1.94 \times 10^9\,\text{m}^3$ during 2012–2016. The net irrigation water consumption in the NW-SDP changed from $1.50 \times 10^9\,\text{m}^3$ during 2001–2005 to $1.83 \times 10^9\,\text{m}^3$ during 2012–2016 and that of the N-HNP changed from $1.78 \times 10^9\,\text{m}^3$ during 2001–2005 to $1.34 \times 10^9\,\text{m}^3$ during 2012–2016.

During the whole winter wheat growing season, approximately four to five times irrigation was applied, with 60–80 mm each, to offset the water deficit of winter wheat growth requirement (Sun et al., 2010). The drops in the groundwater table in the NCP mainly occurred during the winter wheat growing season due to the groundwater extraction for irrigation (Luo et al., 2018). We examined the relationship between net irrigation water use and changes of groundwater levels using the above results (Figure 6.7). The total net irrigation water consumption of winter wheat during 2001–2016 was up to $12.8 \times 10^9\,\text{m}^3$ in the N-HBP and $31.9 \times 10^9\,\text{m}^3$ in the S-HBP. There was a significant negative correlation between the groundwater table and accumulated net irrigation water consumption in the N-HBP and S-HBP (Figure 6.7a and c). In these two sub-regions, with the increase of the accumulated net irrigation water, the groundwater table dropped. This result implied that water consumption for winter wheat production had a significant negative impact on groundwater. But the declining rate of groundwater table had been alleviated since 2001, when the growing area of wheat experienced decreasing trends in both sub-regions (Figure 6.7b and d).

Unlike the Hebei Plain, there was no significant correlation between averaged groundwater table and accumulated net irrigation water consumption in the NW-SDP and N-HNP (Figure 6.8a and c). In these two sub-regions, with the increase of accumulated net irrigation water consumption of winter wheat during 2001–2010, the groundwater table had not shown a significant declining trend. The Yellow River streamflow is the main irrigation water source in these two sub-regions; thus, the groundwater table could be replenished frequently by the surface water irrigation, and had no continuous decreasing trend. Moreover, the total values of net irrigation water consumption during 2001–2016 in the NW-SDP and in the N-HNP were up to $28.5 \times 10^9\,\text{m}^3$. The water withdrawal from the lower reaches of Yellow River started in the 1970s owing to the deficit of rainfall, low runoff, and strong human activities (Chen et al., 2004).

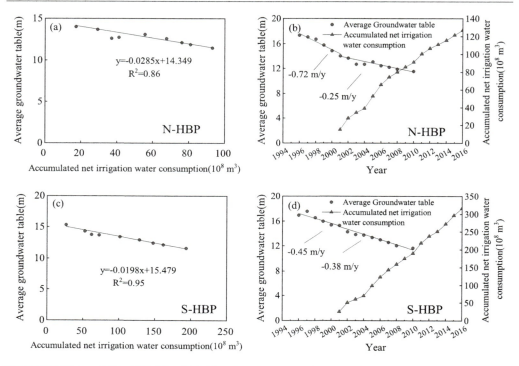

Figure 6.7 The relationship between the average groundwater table and accumulated net irrigation water consumption (a, c), and the change of average groundwater table and accumulated net irrigation water consumption (b, d) in the Northern Hebei Plain and Southern Hebei Plain.

However, in drought years and peak irrigation periods, the irrigation water requirement is not sufficient only from the lower reaches of the Yellow River, and hence groundwater irrigation becomes another option.

6.4 Planting structure adjustment and agricultural water resources saving

Agricultural water in the NCP consumes more than 75% of the portion of groundwater over-exploitation. The amount of shallow groundwater exploitation increased from 3.9 billion m³/year in the 1960s to 7.9 billion m³/year in the 1970s. From 1985 to the end of the 20th century, the average annual groundwater exploitation exceeded 10 billion m³. Out of the pursuit of higher economic benefits and due to the lack of reasonable planning, there is currently a serious surplus of agricultural products in Hebei Province. From 1984 to 2008, the net consumption or exhaustion of groundwater for grain production in Hebei Plain was as much as 139 km³, resulting in an average 7.4 m equivalent drop in groundwater depth in the whole plain and 20 m of groundwater drop in the piedmont plain (Yuan and Shen, 2013). Excessive groundwater consumption has exerted a series of impacts on the ecological environment, social and economic development of the region (Yang et al., 2006). The No. 1 Central Document of China in 2019 and 2020 also mentioned the issues of water saving and agricultural structure adjustment in the NCP, indicating

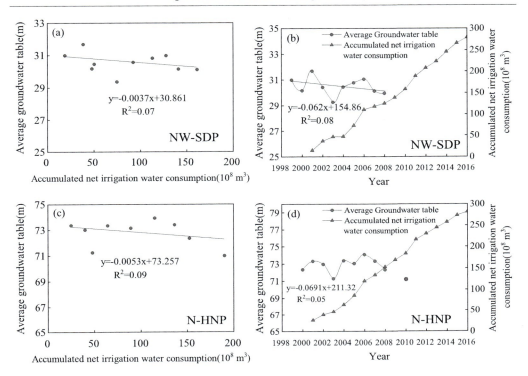

Figure 6.8 The relationship between the average groundwater table and accumulated net irrigation water consumption (a, c), and change of the average groundwater table and accumulated net irrigation water consumption (b, d) in the Northwestern Shandong Plain and Northern Henan Plain.

the urgency and importance of planting structure adjustment in solving the problem of regional groundwater over-exploitation.

Since the implementation of high-efficiency water-saving technology in the 1990s, the water use efficiency has greatly improved in the NCP. However, the total amount of irrigation water consumption has not decreased significantly, and conversely, it has increased due to the expansion of planting scale and irrigation area. This is the rebound effect of "more water saving and more water consumption" (Berbel et al., 2019). At the same time, out of the pursuit of economic benefit, many agricultural producers lacked understanding of market economy and blindly followed others as herding behaviour, resulting in excessive production of agricultural goods. The increase of agricultural input fails to bring an obvious increase in income. Therefore, adjustment and optimization of agricultural planting structure will be of great significance to the sustainable utilization of water resources and sustainable agricultural development.

Constrained by the shortage of water resources and the deterioration of the ecological environment, the agricultural production pattern characterized by "irrigation and high yield" should be changed to a criterion of water use efficiency as soon as possible. As the largest water use sector, agriculture's adjustment or optimization in planting structure could play an important role in decision support for sustainable water resource utilization and improving water use efficiency. In recent decades, the agricultural planting structure in the Beijing-Tianjin-Hebei plain area tended to be simplified, such as winter wheat-summer maize double cropping system being replaced by single maize (Wang et al., 2012; Xu et al., 2019).

From the perspective of water budget, the current cropping pattern with large water deficit needs to be altered with a milder producing intensity to match the rainfall pattern. For example, the average annual precipitation in the NCP is about 500 mm, the ET of wheat-maize system can reach up to 710 mm/year, resulting in a water deficit of more than 200 mm (Shen et al., 2013; Zhang et al., 2018). Optimizing the crop planting structure and limiting the area of high water-consuming crops are also important options to prevent groundwater from continuous depletion in the NCP (Gao et al., 2015; Holst et al., 2014; Hu et al., 2010; Yuan & Shen, 2013). Therefore, adjusting planting patterns and applying field water-saving technologies have the potential to achieve a significant water conservation and high water use efficiency. The facts of groundwater level changes in Hebei Plain present above provide an obvious case to reach a more sustainable cropping pattern/intensity.

6.5 Concluding remarks

The NCP is the second largest plain in China, and it plays an important role in the national grain production mission. Changes in agricultural land use are strongly related to national food security and water resources use. The long-term irrigation has resulted in a rapid decline of the groundwater level in the NCP, forming one of the largest "compound depression cone" areas worldwide. On the other hand, there are some positive signs that happened in past decades due to the changing conditions of agriculture, such as socio-economic environment, agricultural cost and profit; agricultural land use patterns in NCP have undergone a big change, leading to subsequent changes in groundwater use. The case study of spatial and temporal changes in wheat planting area implies that re-planning or optimizing the agricultural planting structure could have a positive effect on groundwater conservation. It is happening naturally, and should be promoted institutionally.

References

Berbel J, Expósito A, Gutiérrez-Martín C, Mateos L. 2019. Effects of the irrigation modernization in Spain 2002–2015. *Water Resources Management*, 33: 1835–1849.

Breiman L, Friedam J H, Olshen, R A, Stone C J. 1984. *Classification and regression trees*. Belmont: Wadsworth International Group, pp. 1–358.

Chen J, Fukushima Y, Tang C, Taniguchi M. 2004. Water environmental problems occurred in the lower reach of the yellow river. *Journal of Japan Society of Hydrology & Water Resources*, 17(5): 1513–1524.

de Oliveira S N, de Carvalho Júnior O A, Gomes R A T, Guimarães R F, McManus C M. 2017. Landscape-fragmentation change due to recent agricultural expansion in the Brazilian Savanna, Western Bahia, Brazil. *Regional Environmental Change*, 17: 411–423.

Fan J, Tian F, Yang Y, Han S, Qiu G. 2010. Quantifying the magnitude of the impact of climate change and human activity on runoff decline in Mian River Basin, China. *Water Science & Technology*, 62(4): 783–791.

Gao B, Ju X, Meng Q, Cui Z, Christie P, Chen X, Zhang F. 2015. The impact of alternative cropping systems on global warming potential, grain yield and groundwater use. *Agriculture, Ecosystems and Environment*, 203: 46–54.

Holst J, Liu W, Zhang Q, Doluschitz R. 2014. Crop evapotranspiration, arable cropping systems and water sustainability in southern Hebei, P.R. China. *Agricultural Water Management*, 141: 47–54.

Hu Y, Moiwo J P, Yang Y, Han S, Yang Y. 2010. Agricultural water-saving and sustainable groundwater management in Shijiazhuang Irrigation District, North China Plain. *Journal of Hydrology*, 393(3–4): 219–232.

Li X, Tong L, Niu J, Kang S, Du T, Li S, Ding R. 2017. Spatio-temporal distribution of irrigation water productivity and its driving factors for cereal crops in Hexi Corridor, Northwest China. *Agricultural Water Management*, 179: 55–63.

Liu J, Kuang W, Zhang Z, Xu X, Qin Y, Ning J, Zhou W, Zhang S, Li R, Yan C, …, Chi W. 2014. Spatiotemporal characteristics, patterns, and causes of land-use changes in China since the late 1980s. *Journal of Geographical Sciences*, 24(2): 195–210.

Luo J, Shen Y, Qi Y, Zhang Y, Xiao D. 2018. Evaluating water conservation effects due to cropping system optimization on the Beijing-Tianjin-Hebei plain, China. *Agricultural Systems*, 159: 32–41.

Pan X. 2015. *Spatio-temporal variation of main crops planting area in the North China Plain using remote sensing data*. Xining, China: Qinghai Normal University.

Qi Y, Luo J, Gao Y, Min L, Han L, Shen Y. 2022. Crop production and agricultural water consumption in the Beijing-Tianjin-Hebei region: history and water-adapting routes. *Chinese Journal of Eco-Agriculture*, 30(5): 713–722 (in Chinese).

Roerink G, Menenti M, Verhoef W. 2000. Reconstructing cloudfree NDVI composites using Fourier analysis of time series. *International Journal of Remote Sensing*, 21(9): 1911–1917.

Shen Y, Zhang Y, R Scanlon B, Lei H, Yang D, Yang F. 2013. Energy/water budgets and productivity of the typical croplands irrigated with groundwater and surface water in the North China Plain. *Agricultural and Forest Meteorology*, 181: 133–142.

Sun H, Shen Y, Yu Q, Flerchinger G, Zhang Y, Liu C, Zhang X. 2010. Effect of precipitation change on water balance and WUE of the winter wheat-summer maize rotation in the North China Plain. *Agricultural Water Management*, 97(8): 1139–1145.

Tang H, Wu W, Yang P, Zhou Q, Chen Z. 2010. Recent progresses in monitoring crop spatial patterns by using remote sensing technologies. *Scientia Agricultura Sinica*, 43(14): 2879–2888 (in Chinese).

Wang J, Wang E, Yang X, Zhang F, Yin H. 2012. Increased yield potential of wheat-maize cropping system in the North China Plain by climate change adaptation. *Climatic Change*, 113: 825–840.

White M, Beurs D, Kirsten M, Didan K, Inouye D W, Richardson A D, Jensen O P, O'Keefe J, Zhang G, Nemani R R. 2009. Intercomparison, interpretation, and assessment of spring phenology in North America estimated from remote sensing for 1982–2006. *Global Change Biology*, 15(10): 2335–2359.

Wu G, Chen S, Su R, Ji M, Li W. 2011. Temporal trend in surface water resources in Tianjin in the Haihe River Basin, China. *Hydrological Processes*, 25(13): 2141–2151.

Wu X, Qi Y, Shen Y, Yang W, Zhang Y, Kondoh A. 2019. Change of winter wheat planting area and its impacts on groundwater depletion in the North China Plain. *Journal of Geographical Sciences*, 29(6): 891–908.

Xu Z, Chen X, Wu S R, Gong M, Du Y, Wang J, Li Y, Liu J. 2019. Spatial-temporal assessment of water footprint, water scarcity and crop water productivity in a major crop production region. *Journal of Cleaner Production*, 224: 375–383.

Yang Y, Watanabe M, Zhang X, Zhang J, Wang Q, Hayashi S. 2006. Optimizing irrigation management for wheat to reduce groundwater depletion in the piedmont region of the Taihang Mountains in the North China Plain. *Agricultural Water Management*, 82(1–2): 25–44.

Yuan Z, Shen Y. 2013. Estimation of agricultural water consumption from meteorological and yield data: a case study of Hebei, North China. *PLoS One*, 8(3): e58685.

Zhang Y, Guo Y, Shen Y, Qi Y, Luo J. 2020. Impact of planting structure changes on agricultural water requirement in North China Plain. *Chinese Journal of Eco-Agriculture*, 28(1): 8–16 (in Chinese).

Zhang Y, Lei H, Zhao W, Shen Y, Xiao D. 2018. Comparison of the water budget for the typical cropland and pear orchard ecosystems in the North China Plain. *Agricultural Water Management*, 198: 53–64.

Zhang Y, Qi Y, Shen Y, Wang H, Pan X. 2019. Mapping the agricultural land use of the North China Plain in 2002 and 2012. *Journal of Geographical Sciences*, 29(6): 909–921.

Chapter 7

Adjustment of field cropping patterns
A better intensity

Yongqing Qi, Jianmei Luo, Yanjun Shen, Dengpan Xiao, and Suying Chen

7.1 Introduction

Agricultural irrigation, accounting for more than 75% of groundwater consumption in the North China Plain (NCP), is the most important factor for serious over-exploitation of groundwater (Sun et al., 2010; Xu et al., 2005). Since the 1970s, the irrigation area of farmland has been greatly expanded in the NCP, and the crop production intensity increased significantly. Gradually, the typical cropping rotation had shifted from three-cropping-in-two-years pattern under rainfed or supplementary irrigation to two-cropping-in-one-year irrigated pattern, forming a winter wheat-summer maize double cropping system (Xiao and Tao, 2014; Zhang et al., 2019). In the summer maize growing season, the total precipitation (300–400 mm) can match the water demand of maize. In the winter wheat growing season, however, the total precipitation (about 120 mm) is too scarce to support the water demand of wheat, whereby high water deficit (about 300 mm) is filled by groundwater irrigation for large-scale wheat production in the NCP (Wu et al., 2019).

Irrigation is the main cause of continuous groundwater over-exploitation in the NCP, which has enabled the region to achieve high and stable agricultural yields over the past several decades, despite incurring significant costs (Hu et al., 2002, 2010). An estimated more than 1 million motor-pumped wells for irrigation have been constructed in the plain of the Beijing-Tianjin-Hebei Region (BTH Region) since the 1970s. The agricultural water consumption accounts for more than 70% of the groundwater exploitation in plain of the BTH Region, far beyond the regional groundwater supporting capacity (Zhang et al., 2012). The net groundwater consumed by the grain production of Hebei Province during 1984–2008 was ~139 km^3, producing a benefit of 190 million tons of grain and causing ~7.4 m decline of the average groundwater level (Yuan and Shen, 2013). Obviously, irrigation-dependent high-yield cropping pattern consumed vast water resources, despite its irreplaceable roles in ensuring regional food security and economic development.

Along with the change in agricultural productivity and irrigation conditions in the last decades of the past century, the double cropping system went through several different stages. In the first half of the 20th century, the NCP was still a typical rainfed farming area, where drylands account for nearly 90% of the total farming area (Zhang, 2012), grain crops mainly included sorghum, millet, wheat and maize (Hui and Kang, 2009), and cropping patterns mainly involved three-cropping-in-two-years, four-cropping-in-three-years and one-cropping-in-one-year patterns (Xu, 1995). The irrigating wells were mainly located in areas with fine storage conditions of shallow groundwater in the piedmont plain of Taihang Mountains, and mainly used for resisting drought on cotton, tobacco and wheat. From the 1950s to the 1970s, the total planting

DOI: 10.1201/9781003221005-10

area remained stable and the grain yield significantly increased; the planting area and yield of wheat and maize have been increasing, and the proportion of irrigated farmland increased from 10.5% to 68.9% with the improvement of irrigation conditions (Luo, 2019), in the process of which traditional shallow wells were gradually replaced by electric-mechanical deep wells; irrigation intention had changed from drought relief to pursuit of high crop production (dry farming changed to irrigated farming), which accelerated improvement of agricultural production capacity. While groundwater over-exploitation became normal, the sustainability of agricultural water resources was gradually diminishing (Pei et al., 2017). From the 1980s to the 2000s, the agricultural production capacity in the NCP was rapidly improved due to the widespread adoption of irrigation, fertilizers, pesticides and high-yield species; the winter wheat-summer maize double cropping system was formed and agricultural production was transformed from "food and clothing" to "well-off" status, but to improve the crop yield remained the first target of regional agricultural production. The grain crops and the cash crops (vegetables and fruits) relied highly on large-scale and high-intensity groundwater irrigation, forming a situation of "water for food" and "water for money", many irrigation equipment (mainly well irrigation) were seen everywhere across the plain, further aggravating the groundwater crisis (Qi et al., 2022). Since the 2010s, the agricultural production and irrigation conditions in the NCP have gone toward a higher level in China; water use efficiency (WUE) and water-saving ratio of irrigation are leading in the country; the total water consumption still exceeds the carrying capacity of sustainable resource utilization despite a continuous decline in agricultural water consumption; with the historical grain shortage problem was addressed, meanwhile the importance and urgency of agricultural production was consequently weakened; agriculture has transitioned into a stage focused on planting structure optimization, water conservation and implementing green solution.

Over the past 40 years, water resources and ecological problems caused by large-scale agricultural irrigation in the NCP, especially the BTH Region, were intensifying (Luo et al., 2018; Qi et al., 2022; Yuan and Shen, 2013). This has formed a dilemma between inter-basin water transfer (need to pay a high cost) and agricultural products outputs (need large consumption of local water resources). In view of the urgent need of a "water-matching" agricultural development pattern, water-saving capacity and planting structure need to be improved (Xiao et al., 2017). The field experiments showed that reducing the proportion of winter wheat in the cropping rotation pattern can significantly decrease irrigation water demand. In addition, through replacing crop types and optimizing growth periods, the yield loss to some extent can be remedied by winter wheat fallow (Guo et al., 2013; Yang et al., 2017; Zhao et al., 2008; Zhang et al., 2011). Effects of different cropping rotation patterns on groundwater have been assessed by crop models (Sun et al., 2015; Xiao et al., 2017). Since 2014, a comprehensive management of groundwater over-exploitation has been implemented in the NCP, which urgently requires a transition to water-matching pattern. In this chapter, based on the results of existing studies, grain production capacity and water consumption characteristics of different cropping rotation patterns and planting structures are analyzed; furthermore, planting intensity and crop patterns that were suitable for water resources carrying capacity are discussed.

7.2 Yield and water consumption characteristics of irrigating cropping patterns in the NCP

We selected the Luancheng Agro-Ecosystem Experimental Station of Chinese Academy of Sciences (Luancheng Station), located in Luancheng County, Hebei Province, 37°53′N, 114°41′E, at an altitude of 50.1 m, as a representative site in the northern part of the NCP. The agricultural production in Luancheng Station presents several ecological characteristics such as intensive farming, high yield, resource constrained, and well irrigated. This region is dominated by a winter wheat-summer maize double cropping system. Lacking precipitation, the agricultural production mainly depends on groundwater irrigation and fertilization, leading to serious over-exploitation of groundwater. Since the 1970s, the groundwater level has dropped to below 40 m at a rate of ~1 m per year on average. In the past decades, long-term cropping rotation tests of different patterns and crop types were conducted in Luancheng Station to assess the water consumption characteristics of diverse cropping rotation patterns; medium and long-term grain production capacity and water resource (involving different patterns and crop types) were assessed by crop model, which was calibrated and validated based on the results of field trials.

7.2.1 Typical irrigating cropping patterns of grain production

7.2.1.1 Winter wheat-summer maize double cropping system

The dominant cropping pattern in the NCP is two-cropping-in-one-year winter wheat-summer maize double cropping system, and the production of winter wheat is critical for maintaining regional food security. Summer is the rainy season in the NCP, about 50% of the rainfall occurs in July and August. This exacerbates the mismatch between monthly rainfall and crop water requirements (Figure 7.1). The precipitation in the summer maize growing season is enough to support the crop water requirement, but the annual water shortage of the double cropping pattern is more than 200 mm. However, winter wheat grows in dry season with little precipitation, highly dependent on irrigation to achieve high yield and therefore being the main factor for groundwater over-exploitation. With the implementation of the groundwater limiting exploitation policy, stabilizing the planting area and yield of winter wheat in over-exploited area remains a pending problem.

ET of the wheat-summer maize double cropping pattern was monitored by an eddy correlation system installed at the integrated observation site of Luancheng Station. Crop production management and fertilize practices of the field trial was carried out in accordance with local farmers' traditions, 2 or 3 times irrigations in winter wheat growing season, and 1-time irrigation for summer maize.

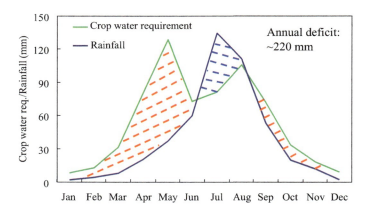

Figure 7.1 The mismatch between monthly rainfall and crop water requirements of the winter wheat-summer maize double cropping pattern in the NCP (1960–2010).

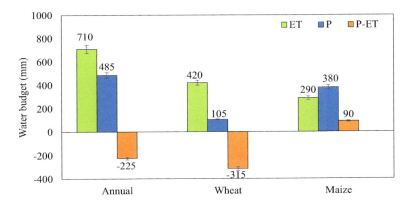

Figure 7.2 The water balance of the winter wheat-summer maize double cropping pattern in Luancheng Station from 2007 to 2016.

The results of water balance calculation from 2007 to 2016 showed that the annual water deficit was 225 mm, while the annual evapotranspiration of 710 mm and precipitation of 485 mm (Figure 7.2). The evapotranspiration of wheat season accounted for about 60% of the total evapotranspiration, so the water deficit mainly occurred in this season up to 315 mm. Due to the shorter growing season of maize, and the higher precipitation compared with wheat season, there is a water surplus of about 90 mm in maize season. Under this water budget level, the annual average yield of two-cropping farmland in the NCP during the same period reached 14,998 kg/ha, among which the average annual yield reached 6,829 kg/ha for the wheat and 8,169 kg/ha for maize.

7.2.1.2 Maize mono-cropping in one year

From 1956 to 2009, the average annual precipitation in the NCP was 538 mm, with less than 400 mm in a dry year and more than 800 mm in a wet year. The rainfall mainly occurred from June to September, highest in July (31% of annual precipitation) and August (25% of annual precipitation), and slight precipitation in the drought season during winter and early spring. This produced a traditional pattern of mono-cropping planting for farmland without irrigation works, where crops are sown at the beginning of the early summer and harvested at the end of the rainy season. To assess the yield ability and water consumption of the maize mono-cropping pattern, soil water transport, deep leakage, and annual water balance were simulated by HYDRUS model. The simulation period was in a maize mono-cropping season from 1976 to 2013, with the fallow period with bare soil (no vegetation and no mulch); maize was sown on May 20 and harvested on September 20, and the crop coefficient and leaf area index (LAI) were similar to that in the two-cropping-in-one-year pattern; crop coefficient with bare soil was assigned a value of 0.4. Based on the characteristics of annual rainfall variation during the simulation period, the irrigated water volume was as follows: 0 mm in a rainy year ($P \geq 571$ mm); 80 mm in an ordinary year (376 mm $< P <$ 571 mm); and 160 mm in a dry year ($P \leq 376$ mm).

Simulation results from the HYDRUS model (Table 7.1) show that average evapotranspiration (ET) of growing season, fallow season and whole year was 391, 81, and 472 mm, respectively. Compared with annual precipitation (498 mm), the annual ET is less of 26 mm.

Table 7.1 Water balance for maize mono-cropping pattern at Luancheng Station

	Annual average	Growing season (5.20–9.20)	Fallow season (9.21–5.19)
Precipitation (mm)	498	384	114
Irrigation (mm)	80	80	0
ET (mm)	472	391	81
Leakage (mm)	106	115	−9

The results show that the maize mono-cropping pattern does not lead to overexploitation of groundwater.

7.2.2 Modeling water balance of different cropping patterns

In order to evaluate the long-term effect of the adjustment of cropping patterns on water consumption and crop yield, the crop model, Agricultural Production Systems sIMulator (APSIM), was applied to simulate the farmland water balance and yield of different cropping patterns. APSIM model is developed by the Australian Agricultural Production Systems Research Group, including multiple modules such as crop growth and development, soil water and soil nitrogen (Holzworth et al., 2014).

Three planting patterns were designed: winter wheat-summer maize (1Y2M) two-cropping-in-one-year pattern, summer maize (SM) or early sown maize (EM) mono-cropping pattern (1Y1M), and winter wheat-SM/fallow-SM (or EM) (WW-SM/F-SM (EM)) three-cropping-in-two-years pattern. The APSIM was employed to simulate the crop growth process under different planting patterns and water management scenarios (M1–M8) from 1981 to 2015 (Xiao et al., 2017). The parameters of different crop growth periods were set based on the long-term crop phenological records of Luancheng Station (Table 7.2). The sowing dates of winter wheat and SM were set as October 5 and June 15 (M1–M6), respectively; the sowing date of early sowing maize in M7 and M8 scenarios was set as May 10, 35 days earlier than that of SM. The model parameters were adjusted and validated by field experiments conducted in 2006–2012 and 2014–2015, respectively.

Table 7.3 shows water balances and yields in different cropping patterns simulated by APSIM: the conventional winter wheat-SM double cropping pattern has the highest water consumption, with a mean ET of 746.2 mm/year under conventional irrigation pattern; the average groundwater overdraft is 258.0 mm/year due to precipitation's inability to meet crop water demand; under full irrigation condition, the ET can reach 876.8 mm/year, potentially resulting in groundwater overdraft of 388.3 mm/year. Transformed from the two-cropping-in-one-year pattern to the SM mono-cropping pattern, the ET under conventional irrigation will be 482.2 mm/year. The rainfall during the maize growing season is slightly higher than crop water requirement, creating a potential recharge to groundwater by 6.1 mm/year. The double amount of irrigation (incremental 80 mm) will lead to an increase in the average annual ET (up to 510.2 mm/year) and an overdraft of groundwater (22.0 mm/year). Under the condition of automatic irrigation (full water supply), the ET and the groundwater overdraft are 536.3 and 48.2 mm/year, respectively. The three-cropping-in-two-years pattern will elevate the ET to 628.6 mm/year and the groundwater overdraft to 140.7 mm/year. In addition, under the fixed irrigation of 180 mm, the average

Table 7.2 Scenario set of different crop planting patterns to evaluate the impacts of alternative cropping patterns on groundwater storage and crop productivity in the NCP

Cropping patterns	Scenario	Irrigation time (t_m) and amount	
		Winter wheat growth period (Oct 5 to Jun 14)	Summer maize growth period (Jun 15 to Oct 4)
1Y2M	M1	80 mm × 4 t_m	40 mm × 2 t_m
	M2	Automated irrigation	Automated irrigation
1Y1M	M3	Fallow	40 mm × 2 t_m
	M4	Fallow	80 mm × 2 t_m
	M5	Fallow	Automated irrigation
2Y3M	M6	80 mm × 4 t_m or fallow	40 mm × 2 t_m
Early maize growth period (May 10 to maturity)			
1Y1M	M7	60 mm × 3 t_m	
	M8	Automated irrigation	

Note: 1Y2M, 1Y1M and 2Y3M denote winter wheat and summer maize two-cropping-in-one-year, early maize or summer maize mono-cropping, and wheat-maize followed by fallow-maize three-cropping-in-two-years, respectively; t_m denotes number of irrigation events; Automated irrigation denotes irrigation determined by APSIM model based on average soil water content to avoid any water stress (above 65% of root-zone field capacity); and M1–M8 denote the different model experiments simulated in the study.

annual ET of the EM mono-cropping pattern is 529.7 mm/year and the groundwater overdraft is 41.6 mm/year. Under the automatic irrigation condition, the average annual ET increases to 556.3 mm/year, and the groundwater overdraft increases to 65.8 mm/year.

Obvious differences of yield existed among different cropping patterns. For the two-cropping-in-one-year pattern under conventional irrigation, the total yield of winter wheat and SM is 14,753 kg/ha, and under automatic irrigation will increase to 19,313 kg/ha. For SM mono-cropping pattern under conventional irrigation, the yield is 7,226 kg/ha; If irrigation is doubled, the yield will be 8,294 kg/ha; under automatic (full) irrigation, the yield reaches 9,791 kg/ha. The yield is 11,336 kg/ha in the three-cropping-in-two-years pattern (M6). For the EM mono-cropping pattern, the yields are 10,809 and 12,242 kg/ha under fixed irrigation (180 mm) and automatic irrigation, respectively.

7.3 Yields and water consumptions of different rainfed wheat-maize rotations

Wheat and maize are the major grain crops in the NCP, accounting for more than 90% of total crop planting area, and therefore, it is important to make clear water consumption and yield response to different wheat and maize rotations. The experiment was conducted at Luancheng Station from 2007 to 2013; the grain yield, water consumption, benefits of different rainfed wheat-maize rotations were analyzed, specially winter wheat-SM two-cropping-in-one-year pattern and spring maize mono-cropping pattern (Chen et al., 2015). The growth period of winter wheat was from mid-early October to the next year of mid-June, and that of SM was from mid-late June to early October; the growth period of spring maize was from mid-late May to mid-September. There's no irrigation treatment in experimental plots from September 2007 to June 2013.

Table 7.3 Water balances and yield characteristics of wheat-maize patterns in different simulation scenarios

Cropping pattern	Scenarios	Precipitation (mm)	Irrigation (mm)	ET (mm)	Leakage (mm)	Over-pumping (mm)	Wheat yield (kg/ha)	Maize yield (kg/ha)	Total yield (kg/ha)
1Y2M	M1	488.0	400.0	746.2	142.0	258.0	7,272.3	7,480.8	14,753.1
	M2	488.0	524.0	876.8	135.7	388.3	9,478.1	9,834.7	19,312.8
1Y1M	M3	488.0	80.0	482.2	86.1	-6.1	-	7,226.0	7,226.0
	M4	488.0	160.0	510.2	138.0	22.0	-	8,294.3	8,294.3
	M5	488.0	168.3	536.3	120.1	48.2	-	9,791.0	9,791.0
2Y3M	M6	488.0	244.5	628.6	103.8	140.7	3,673.7	7,662.4	11,336.1
1Y1M	M7	488.0	180.0	529.7	138.4	41.6	-	10,809.4	10,809.4
	M8	488.0	198.8	556.3	133.0	65.8	-	12,241.5	12,241.5

7.3.1 Yields of different rainfed wheat-maize rotation patterns

The results showed that, under rainfed conditions, the average yield of winter wheat was 3,554.1 kg/ha, with the highest and lowest yields of 4,407.6 and 2,260.1 kg/ha, respectively, and the highest yield was 1.95 times higher than the lowest yield; the average yield of SM was 4,590.5 kg/ha, with the highest and lowest yields of 7,551.6 and 2,129.5 kg/ha, respectively, and the highest yield was 3.55 times higher than the lowest yield; the average yield of spring maize was 6,144.0 kg/ha, with the highest and lowest yields of 8,252.3 and 3,806.0 kg/ha, respectively, and the highest yield was 2.17 times higher than the lowest. Under rainfed conditions, the yield of winter wheat was relatively stable, while the yields of SM and spring maize largely fluctuated over time, especially the SM prone to be influenced by the variable soil moisture at sowing time. There's little rain in many years, so the emergence of maize was delayed, resulting in a significant decrease in maize yield. The annual yields of the winter wheat-SM two-cropping-in-one-year pattern were higher than those of the spring maize mono-cropping pattern, and the yield gap was 2,096.5 kg/ha, more than 34.1%.

7.3.2 Water consumption in different rainfed wheat-maize rotation patterns

The variability of water use characteristics of different crops was determined in actual crop water consumption. Table 7.4 shows the water consumption changes. The water consumption of winter wheat ranged from 192.6 to 290.1 mm (seven-year average is 215.8 mm); the water consumption of SM ranged from 145.2 to 334.0 mm (average is 244.5 mm); and the water consumption of spring maize ranged from 192.8 to 461.2 mm (average is 325.8 mm); the order of water consumption of crops was ranked as: spring maize > SM > winter wheat. The water consumption of the winter wheat-SM pattern was significantly higher than that of the spring maize mono-cropping pattern, with an average gap of 134.4 mm. The water consumption structure of different cropping patterns was significantly different. For the winter wheat, the soil water was reduced during the growing season, when the soil water consumption accounted for 49.3%–73.9% of total water consumption. For the spring maize and SM, the soil water was increased during the growing season, with the transformation of rainfall into soil water. The mean WUE of winter wheat, SM and spring maize was 1.66, 2.04 and 1.98 kg/m^3, respectively.

The average annual precipitation during the experimental period was 480.4 mm. In the rainy, drought and normal years, the precipitation during June-September accounted for 63.1%, 76.0%

Table 7.4 Water consumption characters of winter wheat, summer maize and spring maize under rainfed conditions in the NCP

Crop	SWD (mm)	D (mm)	R (mm)	ET (mm)	GY (kg/ha)	WUE (kg/m^3)	WCC (mm/kg·ha)
Winter wheat	110.93	1.56	106.37	215.75	3,562.1	1.66	0.0622
summer maize	−108.90	1.37	347.31	244.45	4,842.6	2.04	0.0539
Spring maize	−41.16	0.85	367.79	325.79	6,118.0	1.98	0.0547

SWD, soil water depletion; D, soil leakage; R, rainfall; ET, evapotranspiration; GY, grain yield; WUE, water use efficiency; WCC, water consumption coefficient.

and 88.3% of the annual total precipitation, respectively. Winter wheat grows in drought-prone season, and summer (spring) maize grows in rainy and hot seasons. Under rainfed conditions, the correlations between precipitation and crop yields (winter wheat, SM and spring maize) were 0.4, 0.675 and 0.05, respectively, none of which reached significant confidence level. As the impact of previous precipitation on the next crop is concerned, the correlation between precipitation during SM season and the yield of winter wheat is 0.132, and the correlation between precipitation during winter wheat season and the yield of the next SM is 0.559. The above correlation coefficients are statistically insignificant.

7.3.3 Invest-output benefit of different wheat-maize cropping patterns under dry farming conditions

All input and output prices were calculated based on local prices in 2013. The seven-year average ground fertilizer inputs were 450 kg/ha of diamine phosphate and 183.75 kg/ha of urea; the additional fertilizer was urea of 262.5 kg/ha. According to the prices of fertilizers, the cost of fertilizer was 1,915.88 RMB/ha, the additional fertilizer cost was 551.25 RMB/ha and total fertilizer costs were 2,467.13 RMB/ha. For SM and spring maize, average urea inputs were 450 kg/ha and 945 RMB/ha. The total input for winter wheat fertility was 6,513.53 RMB/ha, with total income of 9,261.7 RMB/ha and net income of 2,748.17 RMB/ha; the corresponding results for SM were 3,945, 10,501.6 and 6,556.6 RMB/ha; and for spring maize were 3,945, 13,487.1 and 9,542.1 RMB/ha. The fertilizer input of winter wheat was the highest, 1.65 times higher than that of a total of SM or spring maize; the net income of winter wheat was the lowest, only 41.9% and 28.8% of the net income of SM and spring maize. The net income of the winter wheat-SM two-cropping-in-one-year pattern was 9,262.13 RMB/ha, and it was 279.97 RMB/ha lower than that of the spring maize mono-cropping pattern.

Materials, machinery and labor inputs accounted for 49.3%, 43.8% and 6.9% of the total inputs for winter wheat; seeds, fertilizers and pesticides inputs accounted for 13.89%, 76.77% and 9.34% of the material inputs, respectively. The corresponding results of SM and spring maize were 50.6%, 38.0%, 11.4%, and 37.59%, 47.37%, 15.04%, respectively. Obviously, among the production inputs of the crops, the agricultural materials and machinery inputs occupied a high proportion, and labor inputs accounted for a low proportion. The ratios of output to input of the three crops were 1.42, 2.66 and 3.42, respectively, and winter wheat was the lowest. Therefore, the winter wheat-SM pattern does not have the advantage in terms of economic efficiency.

In this filed trial, the winter wheat-SM pattern had an additional 34.1% of grain yield compared to the spring maize mono-cropping pattern. However, the extra income was negligible as the labor cost was nearly double as in the mono-cropping pattern. Meanwhile, the water consumption in two-cropping-in-one-year pattern was significantly higher than that in spring maize mono-cropping pattern. Furthermore, the winter wheat mainly consumed soil water while in a drought growing season, but the water consumption of SM was supported by rainfall. By an integrated consideration of labor costs, the net incomes, the maize mono-cropping pattern was a better choice under rainfed conditions.

Although rainfed crop production can reduce the consumption of groundwater for irrigation, but there are two obvious disadvantages of rainfed farming compared with irrigating farming: (1) the yield loss of rainfed farming is too large to acceptable for farmers, and (2) the comprehensive cost of rainfed farming is close to irrigating farming. Which determines that rainfed agriculture is not suitable for the grain production of the NCP. Therefore, rainfed production can be used as a groundwater over-exploitation control measure on a small scale, but cannot be implemented on a large scale in the NCP. In recent years, the area of rainfed agriculture has been limited to less than 5% of the total crop sowing area of Hebei Province.

7.4 Optimization scheme of cropping patterns, consider water consumptions and grain yields

7.4.1 Analysis of water consumption and grain yield capacity in different cropping patterns in farmland scale

The water consumption and yield capacity of different cropping patterns in the typical areas of piedmont plain in the NCP (as Luancheng Station) were compared, including the water flux data of eddy covariance system (Liu et al., 2023; Shen et al., 2013), simulating data of crop model (Xiao et al., 2017) and the results of field trials (Yang et al., 2017). The results show that the annual ET varies between 472.0 mm (maize mono-cropping pattern) and 746.2 mm (winter wheat and SM double cropping pattern) and is ranked as follows: two-cropping-in-one-year > three-cropping-in-two-years > mono-cropping ≈ annual precipitation; the average annual yield (excluding cotton mono-cropping pattern) varies between 7,226.0 kg/ha (maize mono-cropping pattern) and 15,738.8 kg/ha (winter wheat and SM double cropping pattern) (Table 7.5).

Figure 7.3 shows the linear relationships of yield and replanting index against ET in different cropping patterns. As replanting index, irrigation amount or growth period increases, both of the crop yield and water consumption will increase accordingly.

Table 7.5 Summary of water consumption characteristics and grain yield capacities in different cropping patterns

Crop types	Cropping patterns	Methods	Water consumption (mm/year)	Over-pumping (mm/year)	Yields (kg/ha/year)
Winter wheat and summer maize	1Y2M	EC	710.0	225.0	15,738.8
Winter wheat and summer maize	1Y2M	Field trials	724.5	252.3	14,116.6
Wheat, maize, cotton and oil seeds	3Y5M	Field trials	647.4	175.2	9,579.4
Wheat, maize and oil seeds	2Y3M	Field trials	615.0	142.8	9,631.0
Wheat, maize, cotton and sweet potato	4Y5M	Field trials	560.6	88.4	8,088.3
Cotton	1Y1M	Field trials	522.5	50.3	4,190.7
Maize	1Y1M	HYDRUS	472.0	−26.0	8,780.0
Winter wheat and summer maize	1Y2M	APSIM	746.2	258.0	14,753.1
Maize	1Y1M	APSIM	482.2	−6.1	7,226.0
Winter wheat and summer maize, maize	2Y3M	APSIM	628.6	140.7	11,336.1
Early maize	1Y1M	APSIM	529.7	41.6	10,809.4
Winter wheat and summer maize, early maize	2Y3M	APSIM	638.0	149.8	12,781.3

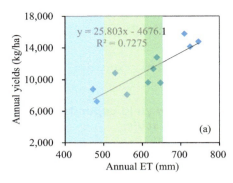 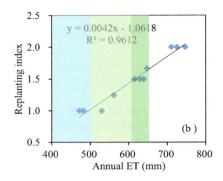

Figure 7.3 Average annual ET in relation to (a) annual yields and (b) replanting index in different cropping patterns.

Results of typical sites showed that the mono-cropping pattern can balance the annual water consumption and precipitation; under the three-cropping-in-two-years pattern, the water deficit at each site ranged from 140.7 to 149.8 mm, indicative of a failure to balance the annual water budget. In the piedmont plain, the average annual precipitation is about 500 mm (In Figure 7.3, marked in blue) and the lateral recharge of groundwater from the adjacent mountainous area is about 113 mm/year (In Figure 7.3, marked in shallow olive) (Chen et al., 2003). Meanwhile, the South-to-North Water Diversion Project (SNWDP) has transferred about 3 billion m³ water per year into the NCP, which is equivalent to 41 mm (In Figure 7.3, marked in green) of water resource as far as the whole region is concerned (Cao et al., 2013). Therefore, in regard to different groundwater recharge pathways (lateral replenishment in mountainous areas, water from the SNWDP and water-saving measures), winter wheat-SM/fallow-EM three-cropping-in-two-years pattern will be a better choice, especially in terms of its benefit on achieving the better relationship of groundwater exploitation and grain yield (Shen et al., 2023).

The above analysis shows: the two-cropping-in-one-year pattern has the highest average annual yield, but fails to ensure groundwater restoration and agricultural sustainability due to high irrigation water consumption; the mono-cropping pattern consumes the least water resource, which can balance the annual water consumption and precipitation recharge, supporting rapid restoration of groundwater resources. However, this pattern significantly reduces the grain yields and threatens the regional food security, and therefore is suitable for being applied in areas with serious groundwater over-exploitation. The three-cropping-in-two-years pattern can balance grain yields and water consumption so that groundwater can achieve a better relation between extraction and recharge.

7.4.2 Optimization scheme of regional cropping system

The water resources are more scarce in the NCP, so the results of field-scale studies in area of the piedmont plain cannot be directly applied to the regional scale. The piedmont plain, where Luancheng Station is located, is an area with high groundwater recharge conditions, but the precipitation is the dominant renewable water resource in other areas in the NCP. Luo et al (2018) developed a framework that integrated crop models, remote sensing ET, and local field trials to optimize crop rotation systems in the BTH region.

In this study, crop coefficient and vegetation remote sensing data were used to estimate water consumptions for crop rotation systems of winter wheat and summer maize on the plain area of the BTH region from 2009 to 2013. For early sowing maize, water consumption and yield were based on the results from the APSIM model, which had been calibrated and validated using field trial data. The effects on groundwater conservation and grain yield were estimated at the regional scale for the conventional winter wheat-summer maize double cropping pattern and three alternative cropping patterns: three-cropping-in-two-years pattern (1st year: winter wheat-summer maize; 2nd year: early sowing maize), four-cropping-in-three-years pattern (1st year: winter wheat-summer maize; the next two years: early sowing maize), as well as a continuous early sowing maize mono-cropping pattern.

There are significant differences in water consumption and grain yield between different cropping patterns. Relative to winter wheat-summer maize double cropping pattern, the water savings in each alternative cropping patterns exceeded 50% of the groundwater overexploitation on the BTH plain. The water consumption of three-cropping-in-two-years pattern, four-cropping-in-three-years pattern, and maize mono-cropping pattern decreased by 14%, 19% and 29% over that of winter wheat-summer maize double cropping pattern (about 740 mm), respectively. The economic water use efficiencies (WUEe) of the alternative cropping patterns were 7% – 16% greater than that of winter wheat-summer maize double cropping pattern.

The groundwater affections were estimated according to the water consumptions of different cropping patterns. The regional groundwater overdraft of winter wheat-summer maize double cropping pattern was 3.83 billion m^3/a (160 mm), and the alternative cropping patterns were more sustainable. The annual average groundwater overdraft amounts were 1.27 billion m^3/a (53 mm) and 0.42 billion m^3/a (17 mm) for the three-cropping-in-two-years pattern and four-cropping-in-three-years pattern, which consumed 2.56 and 3.41 billion m^3/a (or 67% and 89%) less groundwater than double cropping pattern, respectively. For maize mono-cropping pattern, however, the net groundwater overdraft amount was negative, indicating 1.28 billion m^3/a (54 mm) of recharge to groundwater (Figure 7.4). The groundwater drop was mitigated in the alternative cropping patterns. The annual average declines in groundwater by the three-cropping-in-two-years pattern and four-cropping-in-three-years pattern were reduced to 0.21 and 0.08 m/a, respectively, relative to double cropping pattern (0.59 m/a). There was no

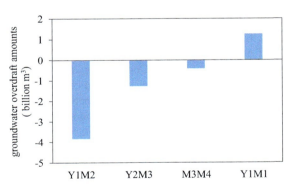

Figure 7.4 Annual net groundwater overdraft amounts for each cropping patterns on the BTH region. (Y1M2: winter wheat-summer maize two-cropping-in-one-year pattern, Y2M3: three-cropping-in-two-years pattern, Y3M4: four-cropping-in-three-years pattern, Y1M1: maize mono-cropping pattern).

groundwater drop, but rather a groundwater rise of 0.15 m in the maize mono-cropping pattern. This is an important reference when regulating grain production policies under the constraint of water resources.

In comparison with yield of winter wheat-summer maize double cropping pattern (29.8 billion kg/a), the total yield was reduced by 2.28, 3.04 and 4.56 billion kg/a (or 7.6%, 10.2% and 15.3%) for three-cropping-in-two-years pattern, four-cropping-in-three-years pattern, and maize mono-cropping pattern, respectively, which accounted for less than 15% of the regional total or less than 1% of the national total. It is important to note that the yield loss of the alternative cropping systems was below the net grain export amount in Hebei province, so the grain supply and demand could be roughly balanced.

In the regional scale of the BTH plain, three-cropping-in-two-years pattern is a preferred alternative cropping system over winter wheat-summer maize double cropping pattern in the near future because of the advantages of high WUEe, economic return, mitigated groundwater level decline and balance between wheat supply and demand on the BTH region. The maize mono-cropping pattern, which had the least water use and highest WUEe and economic return, is an option for the most serious groundwater overdraft area despite a contradiction between the traditional dietary habit and complete abandonment of wheat. The four-cropping-in-three-years pattern is recommended as the transitional cropping system between three-cropping-in-two-years pattern and maize mono-cropping pattern.

7.5 Conclusions

Since the 1970s, in order to meet the increasing food demand in the NCP, the regional crop production has rapidly changed from the traditional three-cropping-in-two-years semi-dry pattern to the wheat-maize two-cropping-in-one-year irrigated high-yield pattern, by investing more irrigation, fertilizers, pesticides and high-yield species. The water shortage and ecological problems caused by agricultural groundwater over-exploitation are increasingly prominent. Therefore, a water-matching, or a water-adaptive crop production pattern should be urgently constructed. This chapter depicted relevant studies and field trials in Luancheng Station, analyzing the grain production capacity and water consumption characteristics of different cropping patterns and crop structures, and discussed the optimal cropping pattern to match water resources.

The water balance, irrigation requirement and grain yield capacity of different cropping patterns were evaluated. Under irrigation conditions, the water consumption of different cropping patterns is ranked as: two-cropping-in-one-year > three-cropping-in-two-years > one-cropping-in-one-year ≈ annual precipitation. Considering various types of supplementary water resources, three-cropping-in-two-years pattern is a good choice of water-adaptive agricultural development, which can achieve a better relationship between groundwater exploitation and recharge, especially in the piedmont plain. The maize mono-cropping pattern or rainfed production can be used as a groundwater over-exploitation control measure on a small scale for the most serious groundwater overdraft area, which had the least water use and highest WUE, but cannot be implemented on a large scale in the NCP. In the process of planting structure optimization, not only the water consumption intensities should be considered, but also the input-output of different crop rotations should be considered. The economic efficiency of farmers is the basis for the effective promotion of regional optimization of the planting structure. On the regional scale of the NCP, three-cropping-in-two-years pattern is a preferred alternative cropping system over winter wheat-summer maize double cropping pattern in the near future because of the advantages of high WUE, economic return, mitigated groundwater level decline and balance between wheat supply and demand on the NCP. The four-cropping-in-three-years pattern is recommended as the transitional cropping system between three-cropping-in-two-years pattern and maize mono-cropping pattern, It is more suitable for balancing regional crop water consumption and grain production on a long time scale.

References

Cao G, Zheng C, Scanlon B R, Liu J, Li W, 2013. Use of flow modeling to assess sustainability of groundwater resources in the North China Plain. *Water Resources Research*, 49(1): 159–175.

Chen J, Tang C, Shen Y, Sakura A, Kondoh A, Shimada J. 2003. Use of water balance calculation and tritium to examine the dropdown of groundwater table in the piedmont of the North China Plain (NCP). *Environmental Geology*, 44: 564–571.

Chen S, Zhang X, Shao L, Sun H, Niu J. 2015. A comparative study of yield, cost-benefit and water use efficiency between monoculture of spring maize and double crops of wheat-maize under rain-fed condition in the North China Plain. *Chinese Journal of Eco-Agriculture*, 23(5): 535–543 (in Chinese).

Guo B, Tao H, Wang P, Knorzer H, Claupein W. 2013. Water utilization of different cropping production systems in North China Plain. *Journal of China Agricultural University*, 18(1): 53–60 (in Chinese).

Holzworth D P, Huth N I, Devoil P G, Zurcher E J, Herrmann N I, McLean G, Chenu K, Oosterom E J, Snow V, …, Keating B A. 2014. APSIM-evolution towards a new generation of agricultural systems simulation. *Environmental Modelling and Software*, 62: 327–350.

Hu C, Zhang X, Cheng Y, Pei D. 2002. An analysis on dynamics of water table and overdraft of groundwater in the Piedmont of Mt. Taihang. *System Sciences and Comprehensive Studies in Agriculture*, 18(2): 89–91 (in Chinese).

Hu Y, Moiwo J P, Yang Y, Han S, Yang Y. 2010. Agricultural water-saving and sustainable groundwater management in Shijiazhuang irrigation district, North China Plain. *Journal of Hydrology*, 393(3/4): 219–232.

Hui F, Kang G. 2009. Study on wheat production and peasants' lives in North China during the Republic of China. *Agricultural History of China*, 28(2): 101–111 (in Chinese).

Liu F, Shen Y, Cao J, Zhang Y. 2023. A dataset of water, heat, and carbon fluxes over the winter wheat summer maize croplands in Luancheng during 2013–2017. *Science Data Bank*, 8(2). DOI: 10.11922/11-6035.csd.2023.0031.zh.

Luo J. 2019. *Evaluating water saving effects due to planting structure optimization in the Beijing-Tianjin-Hebei Plain*. Beijing, China: University of Chinese Academy of Sciences (in Chinese).

Luo J, Shen Y, Qi Y, Zhang Y, Xiao D. 2018. Evaluating water conservation effects due to cropping system optimization on the Beijing-Tianjin-Hebei Plain, China. *Agricultural Systems*, 159: 32–41.

Pei H, Min L, Qi Y, Liu X, Jia Y, Shen Y. 2017. Impacts of varied irrigation on field water budgets and crop yields in the North China Plain: rainfed vs. irrigated double cropping system. *Agricultural Water Management*, 190: 42–54.

Qi Y, Luo J, Gao Y, Min L, Han L, Shen Y. 2022. Crop production and agricultural water consumption in the Beijing-Tianjin-Hebei region: history and water-adapting routes. *Chinese Journal of Eco-Agriculture*, 30(5): 713–722 (in Chinese).

Shen Y, Qi Y, Luo J, Zhang Y, Liu C. 2023. The combined pathway to sustainable agricultural water saving and water resources management: an integrated geographical perspective. *Acta Geographica Sinica*, 78(7): 1718–1730 (in Chinese).

Shen Y, Zhang Y, Scanlon B R, Lei H, Yang D, Yang F. 2013. Energy/water budgets and productivity of the typical croplands irrigated with groundwater and surface water in the North China Plain. *Agricultural and Forest Meteorology*, 181: 133–142.

Sun H, Shen Y, Yu Q, Flerchinger G N, Zhang Y, Liu C, Zhang X. 2010. Effect of precipitation change on water balance and WUE of the winter wheat-summer maize rotation in the North China Plain. *Agricultural Water Management*, 97(8): 1139–1145.

Sun H, Zhang X, Wang E, Chen S, Shao L. 2015. Quantifying the impact of irrigation on groundwater reserve and crop production: a case study in the North China Plain. *European Journal of Agronomy*, 70: 48–56.

Wu X, Qi Y, Shen Y, Yang W, Zhang Y, Kondoh A. 2019. Change of winter wheat planting area and its impacts on groundwater depletion in the North China Plain. *Journal of Geographical Sciences*, 29(6): 891–908.

Xiao D, Shen Y, Qi Y, Moiwo J P, Min L, Zhang Y, Guo Y, Pei H. 2017. Impact of alternative cropping systems on groundwater use and grain yields in the North China Plain region. *Agricultural Systems*, 153: 109–117.

Xiao D, Tao F. 2014. Contributions of cultivars, management and climate change to winter wheat yield in the North China Plain in the past three decades. *European Journal of Agronomy*, 52(Part B): 112–122.

Xu X. 1995. The farming system in the North China Plain in modern times. *Modern Chinese History Studies*, 3: 112–131 (in Chinese).

Xu Y, Mo X, Cai Y, Li X. 2005. Analysis on groundwater table drawdown by land use and the quest for sustainable water use in the Hebei Plain in China. *Agricultural Water Management*, 75(1): 38–53.

Yang X, Chen Y, Steenhuis T S, Pacenka S, Cao S, Ma L, Zhang M, Sui P. 2017. Mitigating groundwater depletion in North China Plain with cropping system that alternate deep and shallow rooted crops. *Frontiers in Plant Science*, 8: 980.

Yuan Z, Shen Y. 2013. Estimation of agricultural water consumption from meteorological and yield data: a case study of Hebei, North China. *PLoS One*, 8(3): e58685.

Zhang G, Fei Y, Wang J. 2012. *Research on adaptability of irrigated agriculture and groundwater in North China*. Beijing, China: Science Press (in Chinese).

Zhang M, Sui P, Chen Y, Sun Z, Ma L. 2011. Water consumption characteristics of alternative cropping rotations in the piedmont of Mt. Taihang. *Chinese Agricultural Science Bulletin*, 27(20): 251–257 (in Chinese).

Zhang X. 2012. *A study on the agricultural production materials and living consumption of farmers in North China in the period of the Republic of China*. Nanjing, China: Nanjing Normal University (in Chinese).

Zhang Y, Qi Y, Shen Y, Wang H, Pan X. 2019. Mapping the agricultural land use of the North China Plain in 2002 and 2012. *Journal of Geographical Sciences*, 29(6): 909–921.

Zhao H, Zhang F, Li J, Tang H. 2008. Direction of agricultural development of urban Beijing: single-harvest spring-maize farming method. *Chinese Journal of Eco-Agriculture*, 16(2): 469–474 (in Chinese).

Chapter 8

Optimization of regional cropping structure

A better planning

Jianmei Luo, Yanjun Shen, and Yongqing Qi

8.1 Introduction

Agriculture has long been a major sector for water use; about 70% of fresh water withdrawal (80%–90% of actual water consumption) is used for agricultural irrigation globally (Foley et al., 2011; Siebert et al., 2010; Wada et al., 2012). The North China Plain (NCP) is one of China's major granaries. In this region, agriculture consumes a large amount of water resources for continuously increasing food production. Especially in the Beijing-Tianjin-Hebei (BTH) Plain (located in the northern part of the NCP), agriculture depends highly on groundwater; although the grain production is ensured, it exacerbates the irreconcilable contradiction between water availability and grain production.

Before the 1970s, the NCP region was a dry farming area dominated by wheat and millet, without any problem of groundwater over-exploitation. Since the 1970s, however, planting intensity is continuously increasing in pursuit of high grain yield to combat the hunger in China. With the improvement of irrigation conditions, the planting system has gradually developed from three harvests in two years, four harvests in three years, or one harvest in one year into a high-intensity irrigation agricultural production mode that is dominated by wheat-maize double cropping system a year (Mo et al., 2009; Wang et al., 2012; Xiao et al., 2013).

The Chinese government officially announced that China had basically solved the problem of food and clothing in 1984. From then on, agriculture has undergone a transformation, with increased planting of high-value crops, such as fruits and vegetables, which consume more water than grain crops (Luo et al., 2022; Zhang et al., 2018). Even though the total amount of agricultural water use decreased in the past three decades owing to large improvement in crop water use efficiency, groundwater level continued to show a declining trend (Figure 8.1, Yang et al., 2021). The total water consumption is still higher than the threshold of sustainable use level.

The shallow groundwater level has dropped on average by 5.5 m from 2000 to 2020 in Hebei Plain, with an average annual decrement of 0.5 m (Hebei Water Resources Bulletin, 2000, 2020). It is reported that 139 billion m³ of groundwater has been consumed for grain production in Hebei Plain during 1984–2008 (Figure 8.2, Yuan and Shen, 2013), and 1.75 billion m³ of shallow groundwater was overdrawn from the piedmont plain area from 1993 to 2012 (Zhang et al., 2016). High-intensity cultivation mode and large amount of irrigation has led BTH region to be the most serious groundwater depletion in China (Cao et al., 2013; Feng et al., 2013, 2018). It is estimated that the shallow aquifer under present abstractions could be depleted to its physical limit within the next 80 years in the piedmont plain (Zhang et al., 2016). The BTH region has become one of the most vulnerable areas in China and possibly worldwide (Wang et al., 2015).

DOI: 10.1201/9781003221005-11

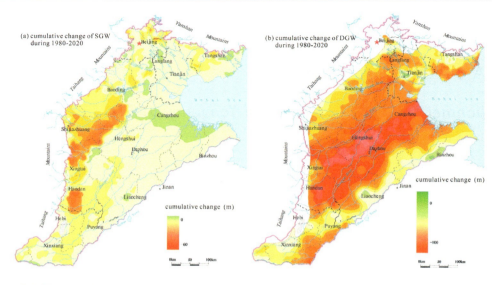

Figure 8.1 The cumulative variation of shallow (a) and deep (b) groundwater level of the NCP in recent 40 years. (Based on Yang et al., 2021.)

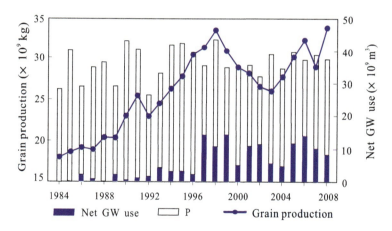

Figure 8.2 Increase of grain production versus net groundwater (GW) consumption in Hebei Plain during 1984–2008. (Modified from Yuan and Shen, 2013.)

Controlling the groundwater over-extraction is becoming an urgent need for sustainable agricultural development.

Previous studies have shown that during the implementation of agricultural water saving, there is a rebound effect that when more water is saved, more water is consumed (Berbel et al., 2019). Moreover, according to the FAO research results based on 20 cases from 14 countries, after the introduction of water-saving technologies, the total irrigation water consumption at the regional scale does not show a decrease, but tends to be increasing (Perry et al., 2017), which is mainly due to the expansion of planting scale and irrigation area (Berbel et al., 2019; Perry et al., 2017; Kang and Tong, 2013). Hence, although the water use efficiency could be improved, the traditional water-saving methods only achieve water saving at the farmland scale rather than the regional scale (Grafton et al., 2018). Under the condition of relatively stable water use efficiency, the scale of crop planting (irrigation) may directly determine the total amount of water consumption. Reducing the scale of planting (irrigation) by adjusting the planting structure is an effective way to mitigate the groundwater depletion.

Currently, China has built a moderately prosperous society in all aspects. Facing severe groundwater over-exploitation, it is necessary to re-examine the issues of agricultural production and water consumption. The first important issue to consider is to explore a reasonable planting scale and structure under water resource constraints, to quantify the water consumption amount of crops, and to arrange suitable agricultural production based on the food requirement in the BTH region. In this sense, we need to confirm the status of crop water consumption and food supply/demand situation, evaluate and optimize the planting structure for a water-suitable planting mode to match with the nature of water resources in this region.

8.2 Agricultural production and future food demand

8.2.1 Development of agricultural production in past half century

There have been various types of crops grown in the BTH region, including wheat, maize, vegetables, fruits, potatoes, beans, cotton, groundnut, millet, sorghum, rice, etc. This situation reflects a self-sufficient farming style in the long pre-mechanization era. After the 1970s, due to technological advances, wheat and maize still maintain their position as the dominant crops, which accounts for 49%–59% of the total sowing area, and 63%–87% of the total production of grain crops, but the planting scale of millet continues to decline, resulting in a significant downgrade in the status of grain production. Millet and sorghum were the crops second only to wheat and maize in the early 1980s, while their growing scale decreased gradually in the near later stages, replaced by vegetables and fruits. Currently, wheat, maize, fruits, and vegetables are the four major crop types in the region, and the total planting area accounts for 84% of the regional total (Figures 8.3 and 8.4).

Even if the structure of crops changed a lot in past decades, the total planting area has not changed much, and the sowing scale of grain crops has decreased by 20% in the BTH region from 1980 to 2016 (Figure 8.4). The planting area of cotton fluctuates greatly, showing a decreasing trend overall, with a decrease of 32%, while that of vegetables, fruits, groundnut, and melon increased by 358%, 422%, 31% and 130%, respectively. The change in agricultural production is a process of the increase of crops with high water consumption, such as vegetables, fruits, melon, as well as the reduction of drought-resistant coarse crops, such as millet, beans, and sorghum.

The production of agricultural products has significantly increased when comparing the two periods of 1981–1985 and 2011–2015 in the BTH region. The yield of grain crops increased from 21.36 to 35.64 million tons, a relative increase of 67% (Figure 8.3). Wheat and maize account for 92% of grain production. Their production has increased by one and two times, respectively. The vegetable production increased from 10.94 to 85.74 million tons. As for fruit, the production increased from 2.06 to 20.71 million tons, with an increase of more than nine times. In the past 35 years, the production of wheat, maize, fruits, and vegetables has increased significantly, reaching an unprecedented high intensity of output in the BTH region.

There are various types of agricultural products in the BTH region. From a perspective of the total amount of agricultural products, the BTH region plays an important role in China and even

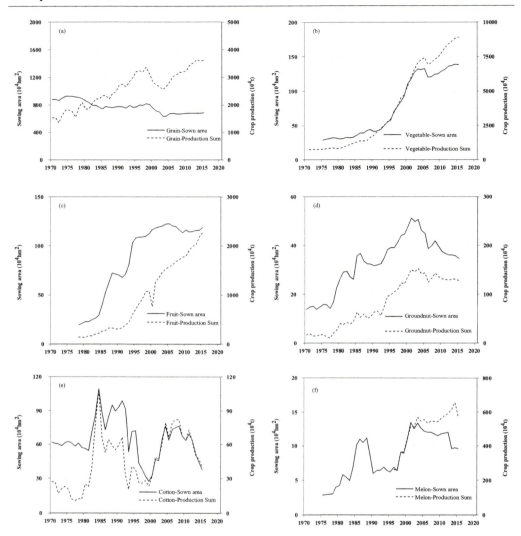

Figure 8.3 Sowing area and yield of crops in the BTH region. a-grain, b-vegetable, c-fruit, d-groundnut, e-cotton, f-mellon.

globally. According to the data from the China Rural Statistical Yearbook and the Hebei Rural Statistical Yearbook, the average annual production of wheat, vegetables, fruits, cotton, meat, poultry eggs, and milk and dairy products in the BTH region ranks at least among the top six in China, with productions of 15.03, 88.56, 20.71, 0.54, 5.35, 19.48, and 6.08 million tons per year, respectively, accounting for 12%, 11%, 8%, 9%, 6%, 13%, and 16% of the corresponding total production in the country (2011–2015). As for the four main crops of wheat, maize, vegetables, and fruits, their annual total production accounts for 6%–12% of the national total and 2%–5% of the global total (Table 8.1), respectively. The BTH region is an important exporting area for agricultural products in China and even the world.

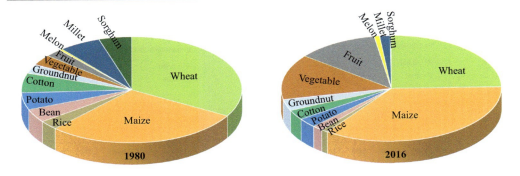

Figure 8.4 Planting structure of crops in the BTH region 1980 (a), 2016 (b).

Table 8.1 Production of crops and the importance of that for the BTH region, 2011–2015

Crop	World	China	BTH	BTHP	China/World	BTHP/China	BTHP/World
	$10^8 t$	$10^8 t$	$10^8 t$	$10^8 t$	%	% (Location order)	%
Wheat	6.92	1.22	0.15	0.14	17.4 (1st)	11.6 (4th)	2.0
Maize	9.30	2.06	0.18	0.17	21.7 (2nd)	8.6 (6th)	1.9
Fruits	6.44	2.45	0.21	0.15	36.2 (1st)	6.1 (6th)	2.3
Vegetables	11.46	7.20	0.86	0.62	62.8 (1st)	8.6 (3rd)	5.4

Notes: BTH, Beijing-Tianjin-Hebei; BTHP, BTH plain.

8.2.2 Estimation of the food demand in 2030

The food demand for the future is an important basis for regulating or re-planning the planting structure. In this sense, food supply is from the agricultural production and food requirement is estimated by the product of population and per capita food consumption. The current food supply is more than the food requirement for most of the agricultural products, for example, the surpluses of fruits, vegetables, eggs, milk, and aquatic products account for more than 50% of the food supply (Luo et al., 2022).

Future food demand is predicted by recommended daily energy intake amount per capita, food energy content, food structure, population, and its gender/age structures (Luo, 2019). The future population is predicted by using two methods, a linear prediction model and a logistic growth model, based on the population census from 1954 to 2016 in the BTH region, which are shown below:

$$Y = 2.2889/\left(1 + 4.01673E + 18e^{-0.0212t}\right) \tag{8.1}$$

$$Y = 0.010695t - 204.7871 \tag{8.2}$$

For the logistic prediction model (Equation 8.1) and linear prediction model (Equation 8.2), the coefficients of determination R^2 are 0.856 and 0.987, respectively, and are significantly correlated at the 0.01 level. The results show that the population would be 126 million (logistic) and 123 million (linear) in the BTH region in 2030 (Figure 8.5). If the National Development

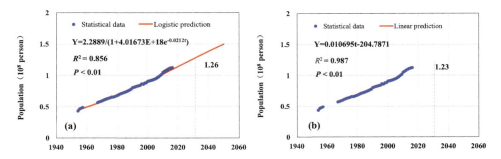

Figure 8.5 The population estimated by the logistic (left panel) and linear (right panel) prediction model in the BTH region in 2030.

Strategy of Xiongan New District is considered, another 1 million people would be added, and the population with a high growth level would be 127 million. The urbanization rate would increase to 0.75 in 2030 (Guangming Daily, 2014). In this sense, high population growth is used. Then, the urban and rural populations would be 95.25 and 31.75 million, respectively, in the BTH region by 2030. Compared with the current population, that is, 110 million in 2020, the population growth will rise 17 million in 2030.

The per capita food demand is estimated using the recommended energy intake amount, the energy consumption structure, and the energy content of crops. The per capita energy intake amount is predicted to be 2,165 and 2,017 kcal/capita/day for urban and rural residents, respectively, in the BTH region in the 2030s, which would be 238 and 304 kcal/capita/day (or 10% and 13%) lower, respectively, than those at present. The energy demand of urban residents will be 148 kcal/capita/day (or 7%) higher than that of the rural residents due to the different population structures, that is, age and gender, of them (Figure 8.6).

The energy consumption structure is very different between urban and rural residents. Grain rations (wheat, rice, and maize) will be the main energy sources, accounting for 50% and 57% of the total energy intake of urban and rural residents (Figure 8.7), respectively, followed by oil,

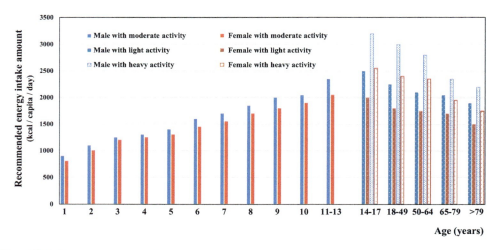

Figure 8.6 The recommended energy intake amount.

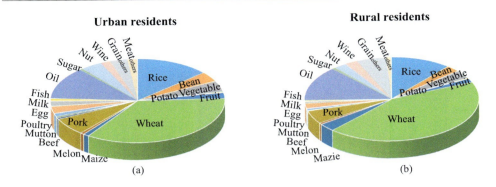

Figure 8.7 The per capita annual energy consumption structure of urban (a) and rural (b) residents in the BTH region.

with the corresponding proportions of 12% and 13%, respectively. As for meat, it shares 11% and 7% of the total energy intake, respectively. Vegetable and fruit together will provide 7% and 5% of the total energy intake for urban and rural residents, respectively. These foods are the main energy sources, accounting for approximately 80% of the energy intake for both the urban and rural residents in the BTH region.

The energy content is different for various foods (Figure 8.8). Oil, nuts, and sugar have high energy contents of 9.00, 5.60, and 3.96 kcal/g, respectively. Grain crops have energy contents from 3.18 to 3.78 kcal/g. The energy content of pork (3.53 kcal/g) is close to that of grain crops, while that of beef, mutton, and poultry is lower, ranging from 2.01 to 2.05 kcal/g. Vegetable and fruit have low energy contents of 0.31 and 0.42 kcal/g, respectively.

On the whole, the per capita food demand will be reduced based on the recommended energy intake amount for a healthy and balanced diet in 2030. As a result, the annual average per capita food demand will be reduced by 10% and 13% for urban and rural residents, respectively.

In future, even though the population will increase by 13–17 million, the current agricultural production will still be higher than the demand in 2030 (Figure 8.9). This can be understood

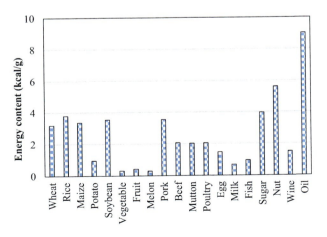

Figure 8.8 The energy content of the dominant foods.

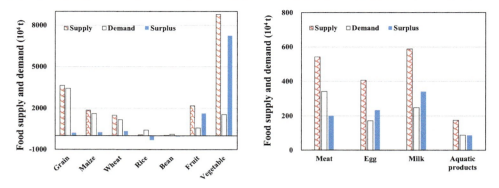

Figure 8.9 Food supply/demand and food surplus of residents in the BTH region in the 2030s. (Data from Luo et al., 2022.) The supply data in this figure used current actual production data in 2016.

by the fact that the changes in population structure, urbanization rate, and improved diet constrain the total food demand not increased linearly as population. Under the scenario of population growth at 127 million, the demand for grain in food crops will be 34.27 million tons (the relative increment will be 3% compared to the current situation); for wheat, demand will be 11.66 million tons, decreased by 1%; and demands for maize, rice, soybean, and potato will be 16.07, 4.10, 1.09, and 0.36 million tons, respectively, with increase of 3%–6%, 4%, 6%, and 3% compared to the current value. The demands for vegetables and fruits will be 15.65 and 5.72 million tons, respectively, increased by 5% and 6%. The demands for meat, eggs, milk, and aquatic products increased by 7%–12%, and will reach 3.42, 1.72, 2.48, and 0.88 million tons, respectively.

Therefore, from the internal perspective of food supply and demand, excessive production is common in the BTH region both at present and in the 2030s. Much of the agricultural exports serve areas outside of the region, and a large amount of virtual water is exported to other areas with the transport of the agricultural products. This does not correspond well with the situation of severe groundwater depletion. On the one hand, the government makes huge efforts including manpower and money to divert water from Yangtze River into NCP to bridge the red water budget, on the other hand, huge amount of virtual water is exported to other regions from BTH region in the form of agricultural product outputs. The blue virtual water exported through agricultural products annually is estimated to be 1.8 times of the water amount allocated to Hebei Province from the SNWDP, that is, 3.47 billion m³. In the future, the planting intensity of crops with high water consumption should be reduced, and an optimized planting scale and structure should be promoted for a better water-grain relationship in the BTH region.

8.3 Water consumption for agricultural production at the regional scale

To clarify the water-saving potential by optimizing planting structure, the water consumption characteristics of each crop and the total water use of the cropping system are quantified in the BTH region, because there are significant differences in irrigation water requirements during the growth period for different crops, which leads to various water consumption among distinct planting structures.

We compared the water requirement, rainfall, and water surplus/deficit for different crops in the BTH region during 2000–2017 (Table 8.2). The annual water requirement of crops includes the water demand in the growing season and fallow period. The annual water demand of the main crops in the BTH plain region is 452–855 mm. Among them, the crops with large annual water demand are vegetables, rice, winter wheat-summer maize, and fruits, more than 702 mm; cotton, potato, and oil crops take the second place, with an annual water demand of 567–644 mm; the average annual water demand of beans, summer maize, and millets is low, between 452 and 530 mm, lower than the mean annual precipitation of 542 mm.

The irrigation water requirement is estimated based on the water deficit of the cropping systems. Negative values indicate a water surplus during the entire year for the cropping system, which is not considered when estimating the total irrigation water requirement.

The annual water deficit of crops in the BTH varies greatly, ranging from −90 to 313 mm for different crops (Table 8.2). The annual water deficit of vegetables, rice, winter wheat-summer maize and fruit trees is 160–313 mm; for cotton, potato, and oil crops, the deficit falls in the range of 25–102 mm; beans, summer maize, and millet have no water deficit in a year, and there is a small amount of water surplus of 12–90 mm.

The planting types with high annual water deficit are vegetables, rice, winter wheat-summer maize, fruits, cotton, and potato, but the planting areas of rice, cotton, and potato are much smaller. Due to the good coupling between crop water demand and precipitation during the growth period of summer maize, there is a small amount of water surplus, and its annual irrigation water demand can be regarded as zero.

Totally, the average water requirement of irrigation is 12.3 billion m³/year in the BTH region. Winter wheat-summer maize, vegetables, and fruits cropping systems consume a large amount of water, with a value of 6.4, 2.8, and 1.9 billion m³/year, respectively, and their water deficits during the growth period account for about 90% of the total water deficit in the farming sector (Luo et al., 2022). Due to the fact that most of the irrigation water in the research area comes from groundwater, the three crops above are also the major crop types using groundwater.

Table 8.2 Mean annual evapotranspiration (ET_c) and water deficit of cropping systems in the BTH region, 2000–2017

Cropping system	ET_c (mm)			Water deficit	
	Whole year	Growing season	Fallow season	Whole year (mm)	Whole year ($10^8 m^3$)
Rice	830	745	85	289	2.9
Winter wheat-summer maize	797	797	0	255	64.2
Maize	503	344	159	−39	−3.5
Millet	452	301	151	−90	−1.4
Bean	530	372	158	−12	−0.2
Potato	627	521	106	86	2.3
Cotton	644	551	93	102	5.4
Groundnut	567	442	125	25	1.2
Vegetables	855	807	48	313	28.4
Fruits	702	702	0	160	18.6
Total					123

Source: Modified from Luo et al. (2022).

The agriculture in the BTH region is characterized by both excessive agricultural production and sever groundwater depletion. During 1984–2008, 139 billion m³ of groundwater has been consumed by grain production, annual average 5.56 billion m³, in Hebei Province (Yuan and Shen, 2013). Besides grain, it is worth mentioning that Hebei is the third and the sixth contributor to the Chinese production of vegetables and fruits in China, and China is the world's largest producer for vegetables and fruits. High water consuming crops are also the crops with large planting scale in the BTH region. Ren (2020) reported that there was 10.17 billion m³ of virtual water in the form of outward transportation of agricultural products in the Hebei Plain region.

Therefore, a large amount of water is consumed during agricultural producing processes, and a huge amount of water has been lost to areas outside of the BTH region with the transport of the agricultural products. An optimized "water-suitable" agricultural cropping system is necessary for water and agriculture sustainability in the BTH region in the new era of social development strategy.

8.4 Optimizing planting structure for different agricultural development scenarios

Optimizing planting structure is an effective way to achieve water-suitable agriculture in the BTH region. On the basis of the carrying capacity of water resources and the food demand of residents in the BTH region, a planting structure optimization model is constructed. The Fast Elite Non Dominant Sorting Genetic Algorithm (NSGA-II) is used to solve the optimization question. Then, we quantify the threshold ranges of water-grain relationship for planting structure optimization scenarios under different groundwater protection objectives. Finally, according to the optimization results of each agricultural development scenario, the total water-saving amount and groundwater resources sustainability are analysed to form a better regional agricultural planning.

8.4.1 Agricultural development scenarios for groundwater conservation

We designed four agricultural development scenarios for considering different groundwater conservation objectives. The first one is business as usual scenario (BAU); the second, optimized planting structure with co-considering the concerns of water consumption, grain production, and farmer's income (OPT); the third, self-sufficiency of major agricultural products in BTH region (SSF); and the fourth, strictly constrained by sustainable groundwater pumping limitation (GWC).

Scenario I (BAU): In this scenario, the planting structure is determined by extrapolating the current situation based on the natural evolution trend of planting structure during the past 15 years. The total sowing area of crops and total water consumption will evolve according to current trends without being affected by groundwater management policies. The results of scenario BAU will provide a reference water-food relationship in 2030.

Scenario II (OPT, coordination of grain crops, cash crops, and water needs): In this scenario, the planting structure is optimized on the basis of BAU, considering the coordination of grain crops (to meet the regional demand of wheat and rice), cash crops (to meet the farmer's needs of increasing income), and water (to meet the groundwater conservation needs). We hope a compromising solution can be found to alleviate the contradiction between water, food, and economy (Luo et al., 2022). The results of scenario OPT could provide information on how much water could be saved or reduced in agricultural consumptive use by comparing with those of BAU, showing the potential of groundwater conservation effect from planting structure optimization.

Scenario III (SSF, self-sufficiency in the main agricultural products): The planting structure of this scenario will be optimized based on the internal food demand or self-sufficiency of main crops (excluding rice) of BTH region without any agricultural products export to other regions. The total water consumption will decrease, while the ecological problem will be mitigated in this scenario. The results of scenario SSF could tell us whether achieving a balanced groundwater budget is feasible by adjusting the agricultural serving objectives to prioritize self-sufficiency, or indicate the size of the remaining gap if balanced groundwater budget attainment proves unattainable.

Scenario IV (GWC, maximum grain production under the constraints of balancing groundwater budget): In this scenario, we assume the adoption of the strictest policies, where groundwater pumping is limited to the red line, signifying zero groundwater level decline. Then the allowable pumpage of groundwater in the region will be used to primarily produce grain for rationing as well as other essential crops required for human dietary needs. The total water consumption will reduce to the balanced level of groundwater extraction and recharge. The results of scenario GWC will help reveal if we could produce human food requirements with the strictest limitation of groundwater governance, and the extent of the food deficit.

For the multi-objective optimization model, the objective functions include meeting minimum water use, maximum economic benefit, and maximum ecological effects. Constraint conditions include cultivated area constraints, water resource constraints, and food supply/demand constraints. Other conditions are that the total water consumption after optimization should be less than the current situation and the ecological benefits should be improved.

8.4.2 Optimized planting structures and water-saving potential

8.4.2.1 Scenario BAU

The optimized planting structures under different development scenarios are shown in Table 8.3. The total sowing area decreases by 2% (Table 8.3). The planting scale of wheat decreases by 11%, while maize and vegetables increase by 20% and 11%, respectively, and the planting scale of fruit changes little in scenario BAU.

Water use of the planting structure for scenario BAU will be 11.3 billion m³/year in the BTH region (Table 8.4). Compared with the current situation, the main crops can save water of

Table 8.3 Results of optimizing planting structure under each scenario based on NSGA-II unit: 10^4 hm²

Scenario	Present	BAU	OPT	SSF	GWC
Grain	665	687	680	622	590
Cotton	53	10	18	23	42
Oil crops	47	33	46	46	46
Vegetables	136	151	116	25	25
Fruits	116	112	95	32	32
Total sowing area	1,017	993	955	748	735

Source: Data from Luo et al. (2022).

Table 8.4 Sowing area, water use, and benefits in each scenario and their changes relative to that of the current planting structure

		Sowing area 10^4 hm^2	Water use 10^8 m^3	Economic benefit 10^8 CNY
Results	Present	1,017	123	2,399
	BAU	993	113	2,391
	OPT	955	105	2,137
	SSF	748	66	1,186
	GWC	735	56	1,177
Changes	BAU	−24	−10	−8
	OPT	−62	−18	−261
	SSF	−269	−56	−1,212
	GWC	−282	−67	−1,222
Change rate	BAU	−2%	−8%	−0.3%
	OPT	−6%	−15%	−11%
	SSF	−26%	−46%	−51%
	GWC	−28%	−54%	−51%

Source: Data from Luo et al. (2022).

1 billion m³/year. Winter wheat-summer maize and cotton will be the major cropping systems contributing to water saving and totally could contribute 1.17 billion m³/year; while water consumption for growing vegetables will increase by 310 million m³/year. This scenario will reduce water consumption by 1 billion m³ annually, but the effect on groundwater conservation will be slight due to the small amount of water reduction and large groundwater depletion (Luo et al., 2022).

8.4.2.2 Scenario OPT

In this scenario, the results show that the planting structure and planting scale of wheat, rice, vegetables, and fruits will be similar to the status quo. The total sowing area will decrease by 380,000 ha compared to scenario BAU. Agricultural water use will be 10.5 billion m³/year, and the water use amount will be reduced by 1.8 billion m³/year relative to the status quo, which will be far less than the regional over-exploitation of 6.7 billion m³/year. We also can know this scenario could create 800 million m³ potential of water saving through planting structure optimization compared to BAU. Therefore, to achieve the coordinated development of water, grain, and economy, water diversion from external basins would be needed to offset the gap of groundwater budget. The amount of extra water sources is around 4.9 billion m³/year to meet the coordinated demands, which is equivalent to 86% of the quantity of water diverted from the SNWDP to the study region (Luo et al., 2022).

8.4.2.3 Scenario SSF

In this scenario, the planting scale of the high water consumption crops, that is, vegetables and fruits, will be reduced to the scale of self-sufficiency. The planting area of main crops will be reduced by 26% relative to the status quo. Grain, vegetables, fruits, and cotton are the main contributors for land reduction. Totally, there are 2.69 million ha of sowing area that is reduced.

The total water use of scenario SSF will be 6.7 billion m³/year with a huge water saving amount of 5.6 billion m³/year in the BTH region. Wheat, vegetables, fruits, and cotton are the main contributors for water use reduction. This will be the maximum water-saving amount when considering the food demand simultaneously. However, there is still about 1.1 billion m³ of water annually in the red, which needs pump from the aquifers. This scenario can achieve the balance of groundwater overdraft in the agricultural sector, but not the total groundwater depletion in the BTH region (Luo et al., 2022).

8.4.2.4 Scenario GWC

The scenario GWC is to try to achieve groundwater neutrality through limiting agricultural water pumping within the extent of rechargeability in the region under the current technology level. Shrinking cropping scale and optimizing their structure are the major regulation measures to achieve this objective. Except for grain crops, the other crops with high water consumption will be reduced to the scale of self-sufficiency. The results show that the total sowing area will be 7.35 million ha with a great reduction of 26.0% compared with scenario BAU.

In this scenario, the total agricultural water use will be 5.6 billion m³/year, reduced by 54% relative to that at present, and the water saving amount is 6.7 billion m³/year. The groundwater level will start to decline. Wheat could not be self-sufficient, although the planting scale of other crops with high water consumption, such as vegetables and fruits, will be reduced to the minimum scale for food self-sufficiency (Luo et al., 2022).

8.5 Effects of optimizing planting structure on water, land, food, and eco-ecologic benefits under different development scenarios

The analysis of the effects of optimizing planting structure on water, land, food, and eco-ecologic benefits can reveal the obstacles that different development scenarios may encounter during the implementation process. There is a nexus between these elements affecting each other. Therefore, some key concerns need to be clarified such as what effects the regulation or optimization of planting structure will have on food security and the economy, and what losses will it cause to groundwater and ecological environment when prioritizing food production? In this sense, comprehensive analysis of the effects of planting structure optimization on water, land, food, and eco-ecologic benefits under different development scenarios needs to be conducted quantitatively for weighing future development strategies.

8.5.1 Groundwater sustainability of different scenarios

8.5.1.1 Water sustainability for the current planting structure

Water sustainability is expressed by the relationship between agricultural water consumption and water resources. Figure 8.10 shows that the average distribution point of agricultural water resource utilization in the BTH region has been located outside the third boundary line; agricultural water use always leads to an unsustainable development of water resources even considering the water from the SNWDP in the BTH region from 2004 to 2015. It is rare for the groundwater to stabilize or even rise to a certain extent, just like in 2012, when water sustainability was high and the total groundwater overdraft amount caused by all the sectors was balanced due to the abundant precipitation. Overall, the average state of agricultural water

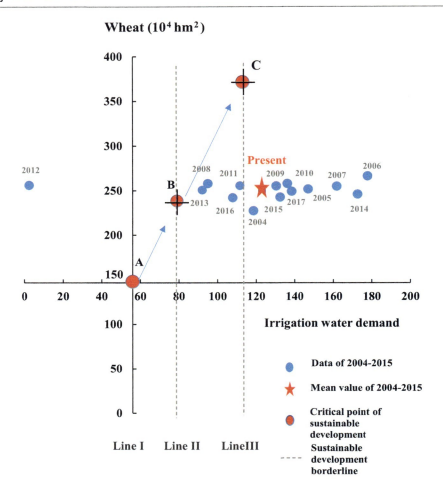

Figure 8.10 Sustainability of crop water use under the present planting structure in BTH. There are three sustainable critical lines for agricultural water use based on the amount of water in the BTH region. It is sustainable within (less than) the critical line and unsustainable otherwise. Lines I and II are the critical lines that can balance the total water overdraft amount (including that caused by agriculture, industry, living, etc.) and the agricultural overdraft amount, respectively, when considering the regional water resources. Line III is the boundary to balance the total overdraft amount considering both the water from regional water resources and from the mid-route of the SNWDP. The coordinate origin is the critical point where the sustainable planting scale of the crops is restricted by regional water resources (not including the imported water from external areas). The irrigation water requirement is estimated based on the water deficit of all cropping systems. There are three points (red and larger dots) of sustainable development on the three sustainable critical lines, respectively, representing the largest planting scale of wheat supported by the water resources in each scenario (on each sustainable critical line).

resources utilization (2004–2015) is unsustainable under the current planting structure even considering the water diversion from the SNWD project.

8.5.1.2 Water sustainability for the optimized planting structures

The distribution points of agricultural water use under scenarios of BAU and OPT are similar to that under the status quo, indicating that the current cropping structure is already a good cropping pattern if water use is not constrained. On the other hand, however, if the water withdrawal is to be limited, there is bound to be some yield or economic loss (Luo et al., 2022). The water use will decrease and the water sustainability will be improved for scenario BAU and OPT, and groundwater sustainability will increase to a safe level if water from SNWD project could be fully used. However, these scenarios are similar to that for the current planting structure, bring about a relative smaller amount of water savings, that is, 1.0 billion and 1.8 billion m³, respectively, and a lower level of water sustainability.

Water sustainability for SSF and GWC is high. SSF could balance the groundwater overdraft amount resulting from agricultural irrigation, but if we want to offset all the groundwater overdraft amount in the study area, about 18% of the water from the SNWD project is required; GWC could balance the total overdraft caused by all the sectors; however, the wheat self-sufficiency rate is only 75% (Luo et al., 2022). Therefore, groundwater over-exploitation is still difficult to be solved completely if we pursue a regionally self-sufficient supply of grain ration. This implies that we have to import grain to reach groundwater balance, or import/divert water from other basins to secure food supply, at current technological condition of water use efficiency (Figure 8.11).

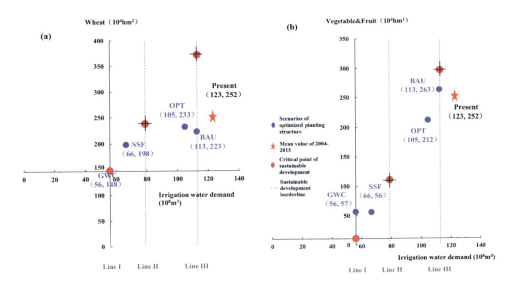

Figure 8.11 Sustainability of crop water use under the optimized planting structure in BTH. (The annotation is as in Figure 8.10.) a-the relationship for wheat sown area and irrigation water demand, b-the relationship for vegetable & fruit sown area and irrigation water demand.

8.5.2 Effects of optimizing planting structure on agricultural land use and food security

"Storing grain in the land" is one of the important agricultural development strategies in China. In this sense, the planting area of the main agricultural crops will be reduced by 0.24–2.82 million hectares, with a rate of 2%–28%. Among them, the scale of wheat decreases by 8%–41%. There will be large farmland that will be fallow under the scenarios of self-sufficiency of agricultural products (SSF) and maximum grain output supported by regional water resources (GWC). This is in line with the national development strategy of fallow and land cultivation (Luo et al., 2022).

Saving water always is accompanied by a certain degree of grain production loss. The total grain output will be changed from −13% to 5% under different scenarios of planting structure optimization, which has little impact on regional grain supply for self-sufficiency. While the planting scale of wheat, the main ration crops in the region, will be 59%–92% of the current scale, it will meet self-sufficiency for wheat in all scenarios except scenario GWC. Totally, to maintain the groundwater neutrality, it will be difficult to meet the wheat demand.

8.5.3 Effects on eco-ecologic benefits under different development scenarios

The ecological benefit is estimated using the economic value of the ecological services of the agro-ecosystems, the negative ecological benefit caused by groundwater pumping, and the value of water saving. Ecological benefit will increase if water saving is implemented through planting structure optimization. The regional ecology and environment will therefore be improved with reducing the planting scale of high water consuming crops. The scenario analysis shows that ecological benefit could increase by one to five times under different scenarios of planting structure optimization (Luo et al., 2022).

The economic benefit will decrease due to reduction in water use and total sowing areas under different development scenarios. Although the direct economic output value of agriculture will drop, the economic benefits could be increased by 8.7–8.9 CNY/m^3 and 90.2–93.1 CNY/m^3 respectively if the saved water in agriculture is used for the secondary and tertiary industries (Zhang and Guo, 2016). It is equivalent to an increase of economic benefits by 8.7–623.77 billion CNY transit from water saving of 1–6.7 billion m^3 in this sense.

Totally, scenario SSF and GWC are at high levels of water sustainability, with large potential for water/land saving and ecological environment improving, while the losses at revenue and food production cannot be ignored. In comparison, scenario BAU and OPT have lower levels of water sustainability, but higher agricultural production and economic benefit relative to scenario SSF and GWC. Here we provide reference thresholds of water/land use and food production for different development scenarios. In practice, planting structure optimization can be determined by the yield and economic objectives of agriculture, and also the willingness to pay for the agricultural production known for "water for food" and "water for money".

8.6 Summary

Groundwater plays an important role in the increased production of agriculture. In the BTH region, a typical characteristic of the highly intensive agricultural production is discernible, leading to a more than ten-fold increase in the output of winter wheat, summer maize, vegetables, and fruit from 1949 to 2015, which has contributed greatly to solving the growing food demand in the BTH region and in China. As a national strategy, the achievement of food self-sufficiency depends largely on the enhancement of cultivation intensity through technology promotion in breeding, irrigation, fertilization, and related agricultural practices. The agricultural system in NCP is highly groundwater dependent today with a total water deficit of 12.3 billion m^3/year.

Optimizing planting structure can alleviate groundwater over-exploitation with the amount of water saved ranging from 1 to 6.7 billion m^3 under different development scenarios, and can reduce the sowing area by 2%–28%, which will not only make water resources more sustainable, ecological environment improved, but also will be in line with the national policy of "Storing grain in the land".

With the implementation of SNWD project and the efforts in suppressing groundwater over-exploitation from 2014, there has been a positive trend in groundwater levels, showing a cease or even recovery of groundwater decline in many locations in the NCP. Actually, it is difficult to stop groundwater decline by only importing water from the SNWD project or by optimizing the planting structure without giving up any grain or economic outputs. To satisfy groundwater sustainability, appropriate importing of wheat for food consumption and importing external water for food production are helpful options for ensuring the coordinated development of regional water and grains in the BTH region.

Within the new era, or post "belly-satisfied" stage, pursuing high agricultural production is no longer the main question of concern, but restoring the severe depleted aquifers and subsequent ecosystems are the major needs. In future, the agriculture in NCP will become more sustainable when new advanced water-saving technologies are developed and applied in practice. In fact, some water-friendly and water smart agricultural technologies are emerging in recent decades, and hopefully are widespread and contribute to sustainable agricultural development.

References

Berbel J, Expósito A, Gutiérrez-Martín C, Mateos L. 2019. Effects of the irrigation modernization in Spain 2002–2015. *Water Resources Management*, 33(5): 1835–1849.

Cao G, Zheng C, Scanlon B R, Liu J, Li W. 2013. Use of flow modeling to assess sustainability of groundwater resources in the North China Plain. *Water Resources Research*, 49(1): 159–175.

Feng W, Shum C K, Zhong M, Pan Y. 2018. Groundwater storage changes in China from satellite gravity: an overview. *Remote Sensing*, 10(5): 674.

Feng W, Zhong M, Lemoine J M, Biancale R, Hsu H T, Xia J. 2013. Evaluation of groundwater depletion in North China using the gravity recovery and climate experiment (GRACE) data and ground-based measurements. *Water Resources Research*, 49(4): 2110–2118.

Foley J A, Ramankutty N, Brauman K A, Cassidy E S, Gerber J S, Johnston M, Mueller N D, O'Connell C, Ray D K, …, Zaks D P M. 2011. Solutions for a cultivated planet. *Nature*, 478(7369): 337–342.

Grafton R Q, Williams J, Perry C J, Molle F, Ringler C, Steduto P, Udall B, Wheeler S A, Wang Y, …, Allen R G. 2018. The paradox of irrigation efficiency. *Science*, 361(6404): 748–750.

Guangming Daily. 2014. Water resources environment in Beijing-Tianjin-Hebei is seriously overloaded. Hydraulic integration urgently needs to break through. https://politics.people.com.cn/n/2014/0921/c70731-25700930.html.

Hebei Water Conservancy Bureau, 2000. Hebei water resources bulletin.

Hebei Water Conservancy Bureau, 2020. Hebei water resources bulletin.

Kang S Z, Tong L. 2013. Deeply understand the evolution law of regional water cycle and establish reasonable scale of oasis irrigation agriculture. *China Water Resources*, 5: 22–25 (in Chinese).

Luo J, Zhang H, Qi Y, Pei H, Shen Y. 2022. Balancing water and food through optimizing the planting structure in the Beijing-Tianjin-Hebei region, China. *Agricultural Water Management*, 262: 107326.

Luo J M. 2019. *Optimization of agricultural planting structure and its water-saving effect in the Beijing-Tianjin - Hebei Plain*. Beijing, China: University of Chinese Academy of Sciences.

Mo X, Liu S, Lin Z, Guo R. 2009. Regional crop yield, water consumption and water use efficiency and their responses to climate change in the North China Plain. *Agriculture, Ecosystems and Environment*, 134(1–2): 67–78.

Perry C, Steduto P, Karajeh F. 2017. *Does improved irrigation technology save water? A review of the evidence*. Cairo, Egypt: Food and Agriculture Organization of the United Nations, p. 42.

Ren D. 2020. *Mutual feedback mechanism and regulation of land-water resources-grain in Beijing-Tianjin-Hebei from the perspective of virtual water*. Beijing, China: University of Chinese Academy of Sciences.

Siebert S, Burke J, Faures J M, Frenken K, Hoogeveen J, Döll P, Portmann F T. 2010. Groundwater use for irrigation: a global inventory. *Hydrology and Earth System Sciences*, 14(10): 1863–1880.

Wada, Y., Beek, L.P.H. Van, Bierkens, M.F.P., Bierkens, M.F.P., 2012. Nonsustainable groundwater sustaining irrigation: A global assessment. *Water Resources Research*, 48, 335–344.

Wang J, Wang E, Yang X, Zhang F, Yin H. 2012. Increased yield potential of wheat-maize cropping system in the North China Plain by climate change adaptation. *Climatic Change*, 113(3–4): 825–840.

Wang X, Li X, Tan M, Xin L. 2015. Remote sensing monitoring of changes in winter wheat area in North China Plain from 2001 to 2011. *Transactions of the Chinese Society of Agricultural Engineering*, 31(8): 190–199.

Xiao D, Tao F, Liu Y, Shi W, Wang M, Liu F, Zhang S, Zhu Z. 2013. Observed changes in winter wheat phenology in the North China Plain for 1981–2009. *International Journal of Biometeorology*, 57(2): 275–285.

Yang H, Cao W, Zhi C, Li Z, Bao X, Ren Y, Liu F, Fan C, Wang S, Wang Y. 2021. Evolution of groundwater level in the North China Plain in the past 40 years and suggestions on its overexploitation treatment. *Geology in China*, 48(4): 1142–1155 (in Chinese).

Yuan Z, Shen Y. 2013. Estimation of agricultural water consumption from meteorological and yield data: a case study of Hebei, North China. *PLoS One*, 8(3): 53–64.

Zhang D, Guo P. 2016. Integrated agriculture water management optimization model for water saving potential analysis. *Agricultural Water Management*, 170: 5–19.

Zhang X, Li R, Kong X. 2016. Estimating spatiotemporal variability and sustainability of shallow groundwater in a well-irrigated plain of the Haihe river basin using SWAT model. *Journal of Hydrology*, 541(Part B): 1221–1240.

Zhang Y, Lei H, Zhao W, Shen Y, Xiao D. 2018. Comparison of the water budget for the typical cropland and pear orchard ecosystems in the North China Plain. *Agricultural Water Management*, 198: 53–64.

Part 4

Better water management for sustainable agriculture

Chapter 9

Managing groundwater quality and quantity of irrigated farmland

Leilei Min, Meiying Liu, Yongqing Qi, and Yanjun Shen

9.1 Introduction

The North China Plain (NCP) is an important grain-producing region in China, contributing to 10.6% of the total grain yield and playing an essential role in economic and social development. There are various agricultural land-use types in the region, such as winter wheat and summer maize, single maize, orchards, vegetables, and cotton, accounting for 54%, 13%, 14%, 7%, and 9% of the farmland, respectively (Zhang et al., 2019). Agricultural production mainly relies on long-term irrigation sourced from groundwater and excessive fertilization (Ju et al., 2009; Pei et al., 2015; Shen and Liu, 2011). Since the 1970s, the farmland irrigation water consumption has sharply increased in the NCP, with more than 70% of the regional water use being allocated for agricultural purposes. Intensive groundwater exploitation has made the NCP one of the most severely overexploited areas worldwide, forming multiple giant groundwater depression cones and causing geological problems such as land subsidence and ground fissures (Xia, 2003; Zhang et al., 1997). Notably, the decline of groundwater levels leads to the formation of a thick vadose zone reaching up to 40 m in the NCP. The surface-applied water moves downward through the vadose zone, serving as a groundwater recharge (Zhang et al., 2009), but the process also carries nitrates and other pollutants downward, leading to a severe threat to groundwater quality (Fei et al., 2014; Wang et al., 2017). Especially, in the context of the policy to reduce the groundwater extraction intensity, the groundwater level has the potential to rebound, which may cause the release of accumulated nitrogen from the vadose zone into the groundwater, thereby contaminating the groundwater. It has been reported that groundwater in some areas has been polluted by non-point source nitrate-nitrogen (Chen et al., 2005; Zhang et al., 2009; Fei et al., 2014; Wang et al., 2017). Previous studies on agricultural water resources primarily concentrated on the root zone (0–2 m) or the aquifer, lacking in-depth knowledge about water and nitrogen processes in the thick vadose zone. In particular, there was limited understanding about the impact of past agricultural cultivation activities and their future transformation patterns on groundwater quantity and quality.

Consequently, ensuring the sustainability of groundwater resources has become a matter of paramount concern, prompting extensive research and intervention efforts (Zheng et al., 2010; Wada et al., 2012). Therefore, the sustainability of groundwater resources must be investigated comprehensively from the perspective of water quality and quantity.

DOI: 10.1201/9781003221005-13

9.2 Groundwater recharge from the perspective of Critical Zone

9.2.1 Concept of the Agricultural Critical Zone

The Critical Zone, spanning from the vegetation canopy to the underground aquifer, is the critical region where the atmosphere, biosphere, pedosphere, lithosphere, and hydrosphere interact with each other, facilitating material migration and energy exchange on the earth's surface (Giardino and Houser, 2015; Li and Ma, 2016). The Critical Zone science emphasizes the use of multidisciplinary theories and methods to investigate the physical, chemical, biological, and ecological processes in the Critical Zone at different time and space scales and analyzes the evolution of the earth's surface system and its controlling mechanisms (Giardino and Houser, 2015).

The Agricultural Critical Zone was a specific type within the broader Critical Zone framework situated in agricultural regions (Lv et al., 2019). In the Agricultural system, the vadose zone is the key region controlling matter, energy, and information flow and transformation. Soil moisture links the physical, chemical, and biological processes of the Agricultural Critical Zone. Vegetation and vadose zones regulate the partition of precipitation and irrigation water, dividing input water into deep drainage (referred to as "seepage" or "groundwater recharge") and evapotranspiration. Meanwhile, water movement is accompanied by contaminant transport (Giardino and Houser, 2015; Nimmo, 2005; Vereecken et al., 2015). Nitrogen, the primary pollutant in water bodies, serves as a crucial nutrient for plant growth but also damages the sustainability of water resources. Therefore, investigating the water and nitrogen transport process in the Agricultural Critical Zone is of great significance for the quality management and sustainability of groundwater resources.

9.2.2 Groundwater recharge in the Agricultural Critical Zone perspective

Vertically, the water cycle in the Agricultural Critical Zone involves hydrological processes such as precipitation, irrigation, evaporation, transpiration, deep drainage, and groundwater extraction (Figure 9.1). In the irrigated farmland with deep groundwater tables, water leaking from the root zone moves downward into the deep vadose zone. After a lag time, it eventually reaches the underlying aquifer, forming the actual groundwater recharge (de Vries and Simmers, 2002).

Climatic conditions, hydrogeological conditions, crop types, and management measures (irrigation, fertilization, etc.) directly affect the field water cycle and redistribution, resulting in variations in evapotranspiration, soil water storage, and deep drainage (Scanlon et al., 2010a; Wang et al., 2013, 2015). Furthermore, vegetation types and management practices have also affected water movement and groundwater recharge. Therefore, to understand the groundwater recharge comprehensively, including its occurrence, processes, quality, and quantity, we should thoroughly consider the crop types (vegetation), soil texture, and water input from the perspective of the Agricultural Critical Zone.

9.2.3 Recharge lag time: celerity vs velocity

There are three significant indicators that reflect water movement in the deep vadose zone: soil water flux, rate of wetting front propagation, and average pore water velocity. The soil water flux in the deep vadose zone is also called Darcy velocity, deep drainage, or potential groundwater

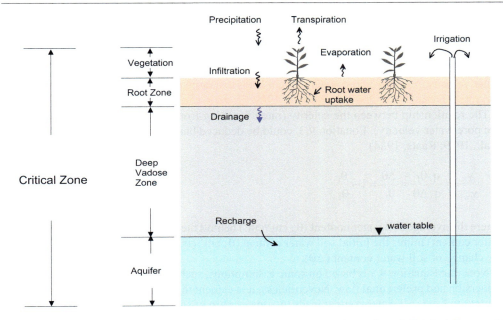

Figure 9.1 Field water cycle and groundwater recharge in the Agricultural Critical Zone with deep water tables.

recharge (Healy and Scanlon, 2010). The rate of wetting front propagation indicates the signal of water pressure propagation, which is also termed "celerity". A faster celerity indicates a quicker response of groundwater table to water input at the ground surface (Scaini et al., 2017). The average pore water velocity reflects the average flow velocity of water in the effective pores of the water-passing section (Zhang et al., 2018), which are widely used to assess the lag time in the impact of land-use changes on groundwater quality (McMahon et al., 2006; Rossman et al., 2014; Scaini et al., 2017; Scanlon et al., 2010b). Quantitative studies of the three rates and an in-depth comprehension of their differences are crucial for understanding the water movement process within the deep vadose zone and analyzing the lag time of groundwater recharge (Jolly et al., 1989; McDonnell and Beven, 2014).

The average pore water velocity (velocity of soil water displacement) can be calculated using Equation 9.1 (Scanlon et al., 2007):

$$v_s = q/\theta_f \tag{9.1}$$

where v_s and q is the average pore water velocity (m/year) and soil water flux (m/year), respectively. θ_f is the average volumetric water content (cm^3/cm^3).

The rate of wetting front propagation (v_{wf}) can be calculated based on Equation 9.2. This equation has been quoted by Warrick (2003), Jury and Horton (2004), and many other scholars (Jolly et al., 1989; Scanlon et al., 2007):

$$v_{wf} = q/\Delta\theta \tag{9.2}$$

where $\Delta\theta$ is the change in volumetric water content (cm^3/cm^3).

Previous studies have shown that the rate of wetting front propagation in the deep vadose zone is generally higher than the average pore water velocity. Under the condition of a constant inflow water flux at the soil surface, the initial water content of the deep vadose zone is the decisive factor for the difference between the wetting front propagation rate and the average pore water velocity. A higher initial water content leads to a more pronounced difference between the two rates (Dahan et al., 2009; Jolly et al., 1989; Scanlon et al., 2007).

The relationship between the celerity (rate of wetting front propagation) and velocity (average pore water velocity), Equation 9.3, could be deduced based on Equations 9.1 and 9.2 (Jolly et al., 1989; Raats, 1984).

$$\frac{v_s}{v_{wf}} = \frac{q/\theta_f}{q/\Delta\theta} = \frac{\Delta\theta}{\theta_f} = 1 - \frac{\theta_i}{\theta_f} \tag{9.3}$$

where the average soil water content (θ_f, cm³/cm³) behind the solute front (also refers to the final water content) minus the initial soil water content (θ_i, cm³/cm³) ahead of the wetting front equals the change of soil water content ($\Delta\theta$).

Note that Equation 9.3 is based on some assumptions, such as the exclusion of hydrodynamic dispersion and preferential flow. Nevertheless, it is evident that a higher ratio of initial soil water content to final soil water content leads to a lower ratio of solute front velocity to wetting front velocity. Field observations have supported the theoretical analysis, showing a close match between the observed values and the theoretical line (Min et al., 2019).

The soil water residence time (t_r, year) could be estimated by Equation 9.4:

$$t_r = L/v_s \tag{9.4}$$

where L is the thickness of deep vadose zone (m).

Similarly, groundwater level response time (t, year) could be estimated by Equation 9.5:

$$t = L/v_{wf} \tag{9.5}$$

Therefore, the term "lag time" may be misleading if the investigating target, soil water displacement or wetting front propagation, was not specified. Sometimes, the term "time lag" was used to refer to soil water residence time (Fenton et al., 2011; McMahon et al., 2006; Scanlon et al., 2010b), whereas some researchers use the term for expressing the water-level response time (Huo et al., 2014; Mattern and Vanclooster, 2010; Scanlon et al., 2010c). Regrettably, little attention has been paid to distinguishing these two terms in the NCP, although some significant results have been reported (Dahan et al., 2009; Jolly et al., 1989; Raats, 1984; Scanlon et al., 2007; Rossman et al., 2014).

9.3 Impact of agricultural activities on groundwater quantity and quality

9.3.1 Status of water and nitrogen application for agricultural management in the NCP

Since the 1980s, agriculture in the NCP has undergone significant transformations. The traditional rain-fed cropping systems (one crop in one year or three crops in two years) have been replaced by the annual double cropping system (winter wheat-summer maize rotation). Irrigation from groundwater pumping is the primary support for stable and high agricultural yield in this region but also the main reason for the continuous overexploitation of groundwater. In the past 20 years, agricultural water consumption has dropped by about 28% (Qi et al., 2022). Fertilizer nitrogen input increased from 200–300 kg/ha/year in the 1980s to 400–600 kg/ha/year in the 2010s. After the 2010s, the nitrogen input in chemical fertilizers slightly decreased to 400–500 kg/ha (Meng et al., 2021). The applied chemical fertilizers will impact or have already affected groundwater quality. According to recent researches, the legacy nitrate-nitrogen accumulated in the deep vadose zone under typical winter wheat and summer maize rotation systems ranges from 118.5 to 6302.8 kg/ha across the NCP. The accumulation is estimated to increase with depth at an average of 157 kg/ha, and the legacy nitrate-nitrogen has entered the aquifer in specific regions (Liu et al., 2022). Therefore, the input of water and fertilizer can impact the groundwater recharge and the quality of recharging water, which will be discussed in the following section.

9.3.2 Impact of field water and nitrogen management practice on groundwater consumption and nitrate leaching

Increasing the water applied at the ground surface undoubtedly leads to more deep drainage. Lu et al. (2021) used the APSIM model (Agricultural Production Systems sIMulator) to simulate the long-term effects of different irrigation schedules for the winter wheat season on nitrogen leaching during the summer maize season from 1981 to 2018. The results showed that the average drainage increased from 1 to 220 mm following rain-fed to five irrigations (with an increase of irrigation from 0 to 600 mm) during the whole year. The average nitrogen leaching corresponding to the drainage changes varied from 0.6 to 90 kg/ha (Figure 9.2). Therefore, reducing the nitrogen leaching by optimizing irrigation management is an effective measure in the NCP.

The field nitrogen management practice impacts evapotranspiration/drainage and nitrate leaching. Nitrogen leaching increased with increasing nitrogen application, especially when the application exceeds 400 kg N/ha/year. Previous studies showed that nitrogen leaching under nitrogen application of 200, 400, and 600 kg N/ha/year was 13.8, 46.3, and 215 kg N/ha/year, respectively (Fang et al., 2008; Wang et al., 2019). The comprehensive results implied that the nitrogen management involving application more than 400 kg N/ha/year is not suitable, because 84% of the additional nitrogen applied would be lost by leaching (Fang et al., 2008). In addition, more nitrogen leaching occurred in the maize season than in the wheat season due to high rainfall in the maize season and high residual soil nitrogen after wheat harvest. In the future, simultaneously optimizing water and nitrogen management and quantitative nitrogen application in crucial growth stages are needed.

The chemical nitrogen application affected evapotranspiration/drainage by affecting above-ground biomass and yield. Studies conducted in the experimental field in Luancheng Station implied that nitrogen application increased evapotranspiration by 9%–27% on average compared to the non-N application treatment in wheat season over a four-year measurement (Liu et al., 2016). Studies based on the chloride mass balance method found that evapotranspiration in the wheat and maize rotation system increased by 2.1%–2.5% under nitrogen application of 400 and 600 kg/ha compared with the non-N application. Correspondingly, deep drainage decreased by 13.2%–16.0% (Wang et al., 2019).

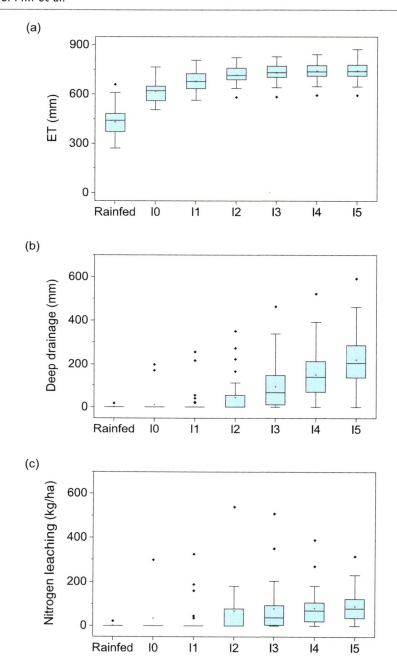

Figure 9.2 Simulated evapotranspiration (a), deep drainage (b), and nitrogen leaching (c) of winter wheat and maize double cropping system at the Luancheng Station from 1981 to 2018. Note that I0 indicates irrigation for germinating, and I1 to I5 indicate irrigation times increased from 1 to 5 after germinating. (Data was obtained from Lu et al. (2021).)

9.3.3 Influence of agricultural land-use types on groundwater recharge and solute transport

The variation in agricultural land use results in differences in irrigation and fertilizer inputs and evapotranspiration, affecting the redistribution and migration of water and nitrogen within the soil. Therefore, the quantity and quality of groundwater recharge also showed significant differences under different agricultural land-use types. According to the studies of Hulugalle et al. (2012) and Scanlon et al. (2010b), the water flux in the deep vadose zone of cotton fields is 22–24 mm/year, and the leaching of nitrate-nitrogen is only 14–26 kg/ha/year. In contrast, orchards and vegetable fields require more water and fertilizer inputs than cotton fields, leading to the water fluxes being 1,100 mm/year in orchards and 132–300 mm/year in vegetable fields. The leaching of nitrate-nitrogen is as high as 80–240 and 346–756 kg/ha/year, respectively (Baram et al., 2016; Onsoy et al., 2005; Soto et al., 2014), which were much higher than those in cotton fields. Scanlon et al. (2010b) have compared water fluxes and salinity in the deep vadose zone of rain-fed and irrigated fields. They found that evapotranspiration was much higher in the irrigated cropland than rain-fed cropland, resulting in almost the same amount of groundwater recharge. The chloride concentration in the deep vadose zone under irrigated cropland is two orders of magnitude higher than that under rain-fed cropland.

The studies conducted by Min et al. (2018, 2019) revealed the groundwater recharge and nitrate leaching in deep vadose zones under four main agricultural land-use types (grain, cotton, fruits, and vegetables) in the NCP. In the piedmont plain, the groundwater recharge under winter wheat and summer maize rotation in one year (WM) with sufficient irrigation was 206 mm/year, and those irrigated with sprinklers were 82 mm/year. For regions involving the single summer maize in one year, fallow land, natural vegetation, vegetables, and orchards, the recharge was 150, 194, 46, 320, and 49 mm/year, respectively (Figure 9.3). In contrast, in the central plain, the recharge was less than that in the piedmont plain, even under the same agricultural land-use types. Under sufficient irrigation conditions, the groundwater recharge was 92.8 mm/year for winter wheat and summer maize rotation. For the single summer maize cultivation, fallow land, vegetable, and cotton fields, it was 51, 85, 256, and 27 mm/year, respectively.

Concerning nitrate-nitrogen leaching, the piston form velocity of downward nitrate transport (average pore water velocity) in the piedmont plain under sufficient irrigated WM is 0.62 m/year. Under sprinkler irrigated WM, single summer maize in one year, fallow land, natural vegetation, vegetables, and orchards, the velocity is 0.29, 0.44, 0.69, 0.60, 2.22, and 0.49 m/year, respectively. In the vegetable growing area, the velocity of downward nitrate transport is greater than the rate of groundwater table decline (0.76 m/year), which has caused the groundwater nitrate to exceed the maximum contaminant level (e.g., 50 mg/L in nitrate) in some areas. Therefore, the conversion of two crops a year to vegetables and fruit trees will affect the quantity and quality of the groundwater to a certain extent.

Conversion of agricultural land cultivation from grain to vegetable could cause great changes in the intensity of irrigation and nitrogen inputs, which potentially poses a great threat to groundwater quality. Thick vadose zone sampling can provide valuable information for evaluating the impact of agricultural land cultivation on groundwater quality. Three adjacent boreholes were collected from grain field of wheat and maize rotation system (G), vegetable fields of six years (V6), and more than 30 years (V30) converted from grain field, respectively (Figure 9.4). The soil water content, and soil chloride and nitrate content were measured. There were no striking differences in the soil water content in the deep vadose zone transitioning from grain field to vegetable fields, except for sporadic layers. The consistency was due to the soil water contents

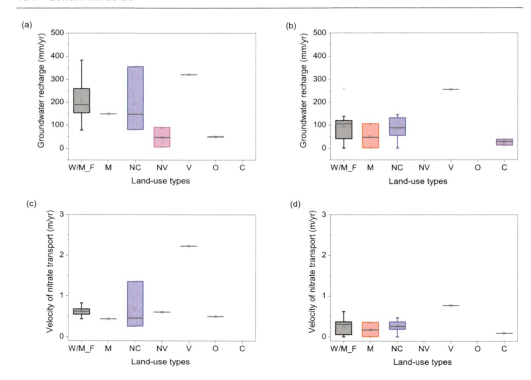

Figure 9.3 Groundwater recharge and velocity of nitrate transport in deep vadose zone under various agricultural land-use types. (a) and (c) indicate values in the piedmont plain, while (b) and (d) represent values in the central plain. Note that W/M_F, M, NC, NV, V, O, and C represent winter wheat and summer maize under flood irrigation, maize, non-cultivation, native vegetation, vegetables, orchards, and cotton, respectively. (Data was obtained from Min et al. (2019).)

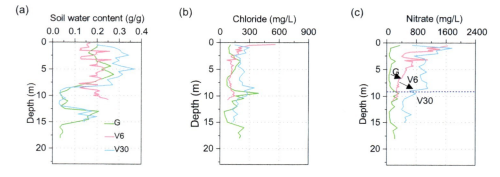

Figure 9.4 Panel (a) soil water content (g/g), (b) chloride (mg/L), and (c) nitrate (mg/L) profiles in the fields converted from grain field of wheat and maize rotation system to vegetable fields in the piedmont plain, the NCP. The symbol G represents wheat and maize rotation system; the symbol V6 and V30 represent vegetable fields of 6 years and more than 30 years converted from grain field, respectively.

in the deep vadose zone of irrigated farmland remaining close to, or slightly over, the field capacity, which was mainly controlled by soil textures. The chloride contents in the vegetable field with a planting history of more than 30 years were more than those in the grain field because of the higher chloride contents in irrigation water. The soil nitrate content changes appeared as sequential breakthroughs across the profile, initiating at the land surface and propagating downwardly. The vegetable field with six years of cultivation showed higher soil nitrate content and it transports to the depth of 9.2 m, with transport velocity of 1.5 m/year. In the vegetable field of more than 30 years, higher soil nitrate content had transported out of the sampling depth. The average soil nitrate content in the vegetable field of more than 30 years was 5.6 times of that in the grain field. These findings can provide mechanistic explanations for analyzing the potential effects of agricultural land use and its transformation on groundwater quantity and quality.

9.3.4 Influence of alternative cropping systems on groundwater consumption and quality

The grain yield, net groundwater consumption, and nitrogen leaching in the past 30 years (1985–2015) were evaluated using the RZWQM model (Liu et al., 2020). The results showed that under two cropping in one year, the average yield in the past 30 years was 11441.5 kg/ha, with the average annual yield of 5245.6 kg/ha for winter wheat and 6195.9 kg/ha for summer maize. The average evapotranspiration was 709 mm, and the annual average nitrate-nitrogen leaching was 55.8 kg/ha for winter wheat and summer maize rotation. After the planting intensity adjustment, for example, changing the winter wheat and summer maize in one year (2M1Y) into three crops in two years (3M2Y), or four crops in three years (4M3Y), or one crop in one year (1M1Y), the simulation results of yield, water consumption, net groundwater consumption, and nitrate leaching amount are shown in Figure 9.5. With the decrease of wheat sowing frequency, water consumption, net groundwater consumption, and nitrogen leaching amount were significantly reduced. Compared with traditional planting, reducing the planting frequency of winter wheat led to a substantial reduction (over 30%) in net groundwater consumption, indicating an effective strategy for decreasing groundwater exploitation and pollution risk (Figure 9.5).

9.3.5 Potential risks to water quality from groundwater table rebound

Since the implementation of the groundwater suppression policy in 2014, some countermeasures have been taken to reduce groundwater extraction in the NCP. Some households and industries switched their water source from groundwater to surface water, supplied through the South-to-North Water Transfer Project. Meanwhile, water-saving measures such as reducing winter wheat planting area and adopting deficit irrigation have been applied to reduce groundwater extraction. These strategies effectively halted the water table decline, supporting the depleted aquifer recovery and ensuring groundwater sustainability. Monitoring data from 559 unconfined wells in the NCP indicate a recent trend of groundwater storage recovery and localized rise in the water table (Zhang et al., 2020). However, given the significant nitrate accumulation in the deep vadose zone of farmland, the rising groundwater level may hasten nitrate infiltration into the aquifer, heightening the risk of groundwater nitrate pollution (Figure 9.6). Therefore, a comprehensive approach to managing both groundwater quality and quantity is imperative for future sustainability.

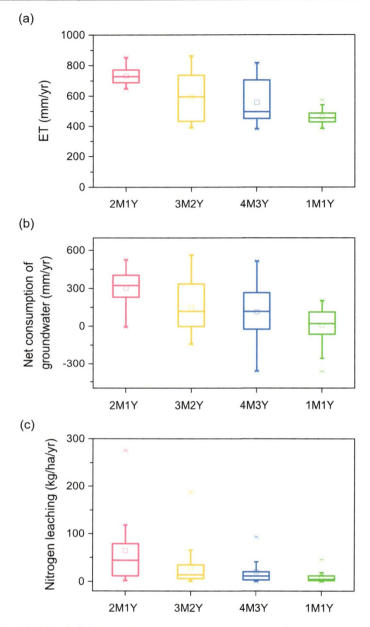

Figure 9.5 The impact of different cropping systems on groundwater consumption and quality of nitrate-nitrogen leaching. (a), (b), and (c) indicate values of evapotranspiration, net consumption of groundwater, and nitrogen leaching, respectively. (Data was obtained from Liu et al. (2020).)

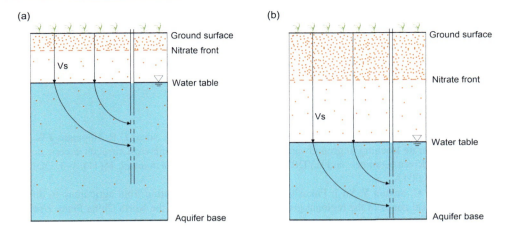

Figure 9.6 The schematic of the threat to groundwater quality. (a) and (b) represent the change of relative positions of water tables and nitrate travel fronts under a groundwater depletion situation. A shift from (a) to (b) indicates that the rate of water table decline is faster than that of nitrate transport (Vs), resulting in an increasing distance between nitrate front and water table. A rise in the water table will shorten the travel path of nitrate, leading to the increased risk of nitrate pollution in groundwater.

9.4 Summary

The historical high-intensity agricultural irrigation in the NCP has led to a continuous decline in groundwater and formed a typical thick vadose zone reaching a depth of more than 40 m. Within the vadose zone, the transport velocity of nitrate is approximately 0.6 m/year, requiring several decades for surface-applied nitrogen fertilizer to seep into the groundwater. Although the groundwater level decline has brought many adverse effects, it has unexpectedly played a positive role in preserving groundwater quality. However, recent concerns have arisen due to climate change and the increased frequency of extreme precipitation events, which heighten the risk of nitrogen leaching. The precipitation timing, frequency, and intensity affect agricultural irrigation demands, subsequently impacting groundwater recharge and nitrogen leaching. Particularly, extreme rainfall events may swiftly transport significant amounts of nitrogen downward. Another concern arises from the conversion of agricultural land types, particularly the shift from grain cultivation to high-water and fertilizer-demanding vegetable cultivation, which imposes greater pressure on both groundwater recharge and water quality. In addition, since the implementation of the policy to reduce the groundwater extraction intensity in 2014, the NCP has observed a trend of groundwater storage recovery and localized groundwater level rebound (Zhang et al., 2020). It indicates the potential groundwater pollution from the accumulated nitrate in the vadose zone.

Based on the national policy, the NCP region continues to bear a substantial agricultural production burden for national food security. This undoubtedly imposes long-term pressure on groundwater resources, with the continuous application of nitrogen fertilizers posing a persistent threat to groundwater quality. Currently, agricultural water productivity has reached a certain level, making further improvements challenging. However, there is still substantial room for conducting risk assessments and optimizing management concerning groundwater quality and quantity. In future researches, it is crucial to analyze the disruptions caused by climate change and agricultural activities on the water and nitrogen transport processes within the thick vadose zone. The analysis might enable a quantitative assessment of their impacts on groundwater quality and quantity, along with the scientific evaluation of region-specific safe groundwater levels. Future researches need to provide theoretical guidance for risk assessment in sustainable groundwater utilization and the formulation of groundwater management policies.

References

Baram S, Couvreur V, Harter T, Read M, Brown P H, Kandelous M, Smart D R, Hopmans J W. 2016. Estimating nitrate leaching to groundwater from orchards: comparing crop nitrogen excess, deep vadose zone data-driven estimates, and HYDRUS modeling. *Vadose Zone Journal*, 15(11): 1–3. DOI: 10.2136/vzj2016.05.0047.

Chen J, Tang C, Sakura Y, Yu J, Fukushima Y. 2005. Nitrate pollution from agriculture in different hydrogeological zones of the regional groundwater flow system in the North China Plain. *Hydrogeology Journal*, 13(3): 481–492.

Dahan O, Talby R, Yechieli Y, Adar E, Lazarovitch N, Enzel Y. 2009. In situ monitoring of water percolation and solute transport using a vadose zone monitoring system. *Vadose Zone Journal*, 8(4): 916–925.

de Vries JJ, Simmers I. 2002. Groundwater recharge: an overview of processes and challenges. *Hydrogeology Journal*, 10: 5–17.

Fang Q, Ma L, Yu Q, Malone R W, Saseendran S A, Ahuja L R. 2008. Modeling nitrogen and water management effects in a wheat-maize double-cropping system. *Journal of Environmental Quality*, 37(6): 2232–2242.

Fei Y, Zhang Z, Guo C, Wang C, Lei T. 2014. Research on the method for evaluation and influence factors identification of regional groundwater quality: a case study of the North China Plain. *Acta Geoscientia Sinica*, 35(2): 131–138 (in Chinese).

Fenton O, Schulte R P, Jordan P, Lalor S T, Richards K G. 2011. Time lag: a methodology for the estimation of vertical and horizontal travel and flushing timescales to nitrate threshold concentrations in Irish aquifers. *Environmental Science & Policy*, 14(4): 419–431.

Giardino J R, Houser C. 2015. Introduction to the critical zone. *Developments in Earth Surface Processes*, 19: 1–13.

Healy R W, Scanlon B R. 2010. *Estimating groundwater recharge*. Cambridge, UK: Cambridge University Press, 245 pp.

Hulugalle N R, Weaver T B, Finlay L A. 2012. Soil water storage, drainage, and leaching in four irrigated cotton-based cropping systems sown in a vertosol with subsoil sodicity. *Soil Research*, 50(8): 652–663.

Huo S, Jin M, Liang X, Lin D. 2014. Changes of vertical groundwater recharge with increase in thickness of vadose zone simulated by one-dimensional variably saturated flow model. *Journal of Earth Science*, 25(6): 1043–1050.

Jolly I D, Cook P G, Allison G B, Hughes M W. 1989. Simultaneous water and solute movement through an unsaturated soil following an increase in recharge. *Journal of Hydrology*, 111 (1–4): 391–396.

Ju X, Xing G, Chen X, Zhang S, Zhang L, Liu X, Cui Z, Bin Y, Christie P, Zhang F. 2009. Reducing environmental risk by improving N management in intensive Chinese agricultural systems. *Proceedings of the National Academy of Sciences of the United States of America*, 106(9): 3041–3046.

Li X Y, Ma Y. 2016. Advances in earth's critical zone science and hydropedology. *Journal of Beijing Normal University (Natural Science)*, 52(6): 731–737 (in Chinese).

Liu L, Hu C, Olesen J E, Ju Z, Zhang X. 2016. Effect of warming and nitrogen addition on evapotranspiration and water use efficiency in a wheat-soybean/fallow rotation from 2010 to 2014. *Climatic Change*, 139: 565–578.

Liu M, Min L, Shen Y, Wu L. 2020. Evaluating the impact of alternative cropping systems on groundwater consumption and nitrate leaching in the piedmont area of the North China Plain. *Agronomy*, 10(11): 1635.

Liu M, Min L, Wu L, Pei H, Shen Y. 2022. Evaluating nitrate transport and accumulation in the deep vadose zone of the intensive agricultural region, North China Plain. *Science of the Total Environment*, 825: 153894.

Lu Y, Wang E, Zhao Z, Liu X. Tian A, Zhang X. 2021. Optimizing irrigation to reduce N leaching and maintain high crop productivity through the manipulation of soil water storage under summer monsoon climate. *Field Crops Research*, 265: 108110.

Lv Y, Hu J, Hu B, Harris P, Wu L, Tong X, Bai Y, Alexis J C. 2019. A framework for the regional critical zone classification: the case of the Chinese Loess Plateau. *National Science Review*, 6(1): 14–18.

Mattern S, Vanclooster M. 2010. Estimating travel time of recharge water through a deep vadose zone using a transfer function model. *Environmental Fluid Mechanics*, 10: 121–135.

McDonnell J J, Beven K. 2014. Debates-the future of hydrological sciences: a (common) path forward? A call to action aimed at understanding velocities, celerities and residence time distributions of the headwater hydrograph. *Water Resources Research*, 50(6): 5342–5350.

McMahon P B, Dennehy K F, Bruce B W, Böhlke J K, Michel R L, Gurdak J J, Hurlbut D B. 2006. Storage and transit time of chemicals in thick unsaturated zones under rangeland and irrigated cropland, High Plains, United States. *Water Resources Research*, 42: W03413.

Meng F, Wang K, Xiao G, Wang K, Hz Z, Zhang H, Xu X, Zhang W, Yang X. 2021. Nitrogen leaching mitigation in fluvo-aquic soil in the North China Plain. *Chinese Journal of Eco-Agriculture*, 29(1): 141–153 (in Chinese).

Min L, Qi Y, Shen Y, Wang P, Wang S, Liu M. 2019. Groundwater recharge under irrigated agro-ecosystems in the North China Plain: from a critical zone perspective. *Journal of Geographical Sciences*, 29(6): 877–890.

Min L, Shen Y, Pei H, Wang P. 2018. Water movement and solute transport in deep vadose zone under four irrigated agricultural land-use types in the North China Plain. *Journal of Hydrology*, 559: 510–522.

Nimmo J R. 2005. Unsaturated zone flow processes. In: Anderson, M G, Bear, J. (Eds.), *Encyclopedia of hydrological sciences*. Chichester, UK: Wiley, pp. 2299–2322.

Onsoy Y S, Harter T, Ginn T R, Horwath W R. 2005. Spatial variability and transport of nitrate in a deep alluvial vadose zone. *Vadose Zone Journal*, 4(1): 41–54.

Pei H, Scanlon B R, Shen Y, Reedy R C, Long D, Liu C. 2015. Impacts of varying agricultural intensification on crop yield and groundwater resources: comparison of the North China Plain and US High Plains. *Environmental Research Letters*, 10(4): 044013.

Qi Y, Luo J, Gao Y, Min L, Han L, Shen Y. 2022. Crop production and agricultural water consumption in Beijing-Tianjin-Hebei Region: history and water-adapting routes. *Chinese Journal of Eco-Agriculture*, 30(1): 1–11 (in Chinese).

Raats P A C. 1984. Tracing parcels of water and solutes in unsaturated zones. In: Yaron B, Dagan G, Goldshmid J (Eds.), *Pollutants in porous media: the unsaturated zone between soil surface and groundwater*. Berlin, Heidelberg: Springer, pp. 4–16.

Rossman N R, Zlotnik V A, Rowe C M, Szilagyi J. 2014. Vadose zone lag time and potential 21st century climate change effects on spatially distributed groundwater recharge in the semi-arid Nebraska Sand Hills. *Journal of Hydrology*, 519: 656–669.

Scaini A, Audebert M, Hissler C, Fenicia F, Gourdol L, Pfister L, Beven K J. 2017. Velocity and celerity dynamics at plot scale inferred from artificial tracing experiments and time-lapse ERT. *Journal of Hydrology*, 546: 28–43.

Scanlon B R, Reedy R C, Gates J B, Gowda P H. 2010a. Impact of agroecosystems on groundwater resources in the Central High Plains, USA. *Agriculture Ecosystems & Environment*, 139(4): 700–713.

Scanlon B R, Gates J B, Reedy R C, Jackson W A, Bordovsky J P. 2010b. Effects of irrigated agroecosystems: 2. Quality of soil water and groundwater in the southern High Plains, Texas. *Water Resources Research*, 46(9): W09538.

Scanlon B R, Reedy R C, Gates J B. 2010c. Effects of irrigated agroecosystems: 1. Quantity of soil water and groundwater in the southern High Plains, Texas. *Water Resources Research*, 46(9): W09537.

Scanlon B R, Reedy R C, Tachovsky J A. 2007. Semiarid unsaturated zone chloride profiles: archives of past land use change impacts on water resources in the southern High Plains, United States. *Water Resources Research*, 43(6): W06423.

Shen Y, Liu C. 2011. Agro-ecosystems water cycles of the typical irrigated farmland in the North China Plain. *Chinese Journal of Eco-Agriculture*, 19(5): 1004–1010 (in Chinese).

Soto F, Gallardo M, Gimenez C, et al. 2014. Simulation of tomato growth, water and N dynamics using the EU-Rotate_N model in Mediterranean greenhouses with drip irrigation and fertigation. *Agricultural Water Management*, 132: 46–59.

Vereecken H, Huisman J A, Hendricks Franssen H J, Bruggemann N, Bogena1 H R, Kollet S, Javaux M, van der Kruk J, Vanderborght J. 2015. Soil hydrology: recent methodological advances, challenges, and perspectives. *Water Resources Research*, 51(4): 2616–2633.

Wada Y, van Beek L P H, Bierkens M F P. 2012. Nonsustainable groundwater sustaining irrigation: a global assessment. *Water Resources Research*, 48: W00L06.

Wang S, Wei S, Liang H, Zheng W, Li X, Hu C, Currell M J, Zhou F, Min L. 2019. Nitrogen stock and leaching rates in a thick vadose zone below areas of long-term nitrogen fertilizer application in the North China Plain: a future groundwater quality threat. *Journal of Hydrology*, 576: 28–40.

Wang S, Zheng W, Currell M, Yang Y, Zhao H, Lv M. 2017. Relationship between land-use and sources and fate of nitrate in groundwater in a typical recharge area of the North China Plain. *Science of the Total Environment*, 609: 607–620.

Wang Y, Hu W, Zhu Y, Shao M A, Xiao S, Zhang C. 2015. Vertical distribution and temporal stability of soil water in 21-m profiles under different land uses on the Loess Plateau in China. *Journal of Hydrology*, 527: 543–554.

Wang Y, Shao M, Liu Z. 2013. Vertical distribution and influencing factors of soil water content within 21-m profile on the Chinese Loess Plateau. *Geoderma*, 193–194: 300–310.

Xia J. 2003. Water cycle and water resources safety in North China: problem and challenges. *Haihe Water Resources*, 4: 1–4 (in Chinese).

Zhang C, Duan Q, Yeh P J F, Pan Y, Gong H, Gong W, Di Z, Lei X, Liao W … Guo X. 2020. The effectiveness of the South-to-North water diversion middle route project on water delivery and groundwater recovery in North China Plain. *Water Resources Research*, 56(10): e2019WR026759.

Zhang R Q, Liang X, Jin M, Wan L, Yu Q C. 2018. *Fundamentals of hydrogeology* (7th Edition). Beijing, China: Geological Press.

Zhang Y, Qi Y, Shen Y, Wang H, Pan X. 2019. Mapping the agricultural land use of the North China Plain in 2002 and 2012. *Journal of Geographical Sciences*, 29(6), 909–921.

Zhang Z, Fei Y, Chen Z et al. 2009. *The investigation and evaluation of sustainable utilization of groundwater resources in North China Plain*. Beijing, China: Geological Press.

Zhang Z, Shi D, Shen Z, Zong Z, Xue Y. 1997. Evolution and development of groundwater environment in North China Plain under human activities. *Acta Geoscientia Sinica*, 18(4): 337–344 (in Chinese).

Zheng C, Liu J, Cao G, Kendy E, Wang H, Jia Y. 2010. Can China cope with its water crisis? Perspectives from the North China Plain. *Groundwater*, 48(3): 350–354.

Chapter 10

Utilization of brackish and other abnormal water in irrigation

Yucui Zhang, Yanjun Shen, and Yongqing Qi

10.1 Utilization of brackish and other abnormal water resources

Brackish/saline water, rainwater/floodwater, and reclaimed water are the three main types of abnormal water. Brackish/saline water generally refers to groundwater with a high degree of salinity. Rainwater and floodwater involve storing and converting rainwater and floodwater into usable water resources, by various engineering techniques and dispatching management methods. Reclaimed water refers to industrial or domestic wastewater which have been properly treated to meet water quality indicators. All these types of abnormal water are important alternative sources for agricultural irrigation.

10.1.1 Utilization of reclaimed, brackish, and other abnormal water resources in agriculture

There is a long history of abnormal water irrigation. Developed countries with advanced agricultural technologies, such as the United States and Israel, are relatively mature in abnormal water irrigation technology and standards. In 2020, the annual utilization of reclaimed water in California reached 2.5 billion m^3, accounting for about 40% of the planned new water resources. Europe has gradually established reclaimed water utilization systems since the 1990s to cope with increasing water demand, and nowadays, about three-fourths of the reclaimed water is used for agricultural irrigation (Bixio et al., 2006). With strict water shortage issue, Israel reclaimed and used about 83% of its total treated sewage in agricultural irrigation, accounting for 40% of the total agricultural irrigation water.

China has adopted wastewater irrigation since the 1950s, and the total wastewater irrigation area has reached 3 million ha now. The annual reclaimed water resources amounted to 36.65 billion m^3 by 2015, of which 11 billion m^3 was used for farmland irrigation. Beijing is one of the cities with the highest utilization rate of reclaimed water in China. It has successively built reclaimed water irrigation districts with an area of more than 40,000 ha by 2010.

As a supplemental water source, brackish water was widely used for irrigation in dry land and saline regions worldwide. Proper use of brackish water for agricultural irrigation is an important way to cope with agricultural droughts and prevent yield reduction. The United States, Australia, Israel, and other countries have established advanced systems for irrigation and utilization of brackish water. In the arid regions of the southwestern United States, underground brackish water is used to irrigate salt-tolerant cash crops such as cotton and alfalfa to achieve good yields. Agricultural production accounts for a high proportion of brackish water irrigation, and brackish water sprinkler and drip irrigation technologies are at the leading level in Israel.

DOI: 10.1201/9781003221005-14

More than 400,000 ha of arable land in the Mediterranean coast of Egypt is irrigated with brackish water, consuming 300–500 million m³ every year.

Underground brackish water resources are widely distributed in North China, Northwest China, and eastern coastal regions, where the burial depth of brackish water is shallow and salinity is always subject, obviously, to rainfall seasonally. The total amount of shallow brackish water resources with a chemical degree of 2–5 g/L is estimated at about 7.5 billion m³ in the North China Plain (NCP), with great potential to be exploited for agricultural irrigation. Since the 1960s and 1970s, China has gradually carried out research and development of brackish water irrigation technologies, and established several good utilization patterns, especially adopting irrigation practices with crop growth requirement and the seasonal variation of water and soil salinity. With the supplement of brackish water irrigation, crop yield in the NCP has been promoted greatly from about 3 t/ha in the beginning of 1980s to more than 12 t/ha today. Even though there are huge contributions from breeding, disease and insect pest protection, and other cultivation technologies, the use of brackish water is the most important driver to improve production.

Rainwater and floodwater can be stored by rivers and wetlands increasing groundwater supply and indirectly making more water resources available for agriculture. Meanwhile, ponds, water cellars, and other facilities also can be used for these water collection and storage. Japan adopts the utilization method of in situ digestion of rainwater and floods, and artificially creates beaches and wetlands on the original channelized rivers, which improves the resource conversion efficiency of rainwater and floods. The United States and Australia collect and store rainwater in urban areas and use it for suburban agriculture, replacing 18%–21% of irrigation water consumption (Lundy et al., 2018). Since the 1980s, UNDP and the World Bank have promoted rainwater storage projects in Africa, using 10–100 m³ storage ponds to collect rainwater and runoff to supplement irrigation or domestic water. About 3.5 million ha of rain-fed arable land in Ethiopia has thus obtained certain irrigation conditions, which has become one of the important measures to alleviate the shortage of food production and domestic water in the dry season (Ngigi et al., 2005). The arid and semi-arid regions of northern China have a long history of agricultural utilization of rainwater and floodwater. In the Qin and Han Dynasties in 2000 BC, ponds and dams were built to collect and store rainwater for irrigation. Since the 1980s, large-scale research, development, and demonstration of rainwater collection and supplementary irrigation technologies have been carried out in Northwest China. Combining rainwater harvesting and water-saving irrigation technology in production has stabilized and increased agricultural production in various places.

Compared with traditional water resources, there are still some obstacles in the agricultural utilization of abnormal water, mainly including enrichment and safety of pollutants and salts in reclaimed water and brackish water, transportation and scheduling of rainwater and floodwater for irrigation, as well as collection, storage, and cost issues for cross-season utilization.

Water quality is the most important factor determining the risk of reclaimed water and brackish water irrigation utilization. To ensure the sustainability of agricultural irrigation, corresponding irrigation technology and supporting crop production technology need to comprehensively consider the following aspects: (1) Long-term brackish water irrigation leads to soil salinization, degradation of land productivity, and saline stress to crops. (2) Water quality standards for reclaimed water irrigation have been established to address concerns such as soil heavy metal accumulation and other pollutants resulting from long-term irrigation practices. (3) Accumulation of heavy metals and pollutants in crops and agricultural products poses significant risks to human health. (4) Reclaimed water and brackish water irrigation can have adverse effects on the saline content in groundwater and the diffusion of pollutants.

In China, the government issued national standards of 'Pollutant Discharge Standards for Urban Sewage Treatment Plants (GB18918-2002)' and 'Water Quality for Farmland Irrigation for Urban Sewage Recycling (GB20922-2007)'. Owing to various factors such as wastewater source, treatment technology, cost, the existing wastewater treatment technology cannot completely remove heavy metals and organic pollutants. Most pollutants do not reflect the apparent pollution characteristics in soil and crops in a short term, so the extended effects of brackish water irrigation on soil salinity and productivity are still the main concern. Soil health is crucial to the sustainable development of agriculture and the environment, and for stable crop production as well.

10.1.2 Utilization of abnormal water resources in the NCP

10.1.2.1 Reclaimed water utilization

Under the situation of water shortage in the NCP, reclaimed water is an important abnormal water source, and its utilization has increased year by year. The total amount of sewage treatment and recycled water utilization in 2015 was 2.4 and 4.1 times of the numbers in 2005, respectively. In 2018, the utilization of reclaimed water in Beijing-Tianjin-Hebei increased to 2.016 billion m^3, equivalent to 6.7 times of the ability in 2005. The utilization of reclaimed water is 1.076 billion m^3 in Beijing, 414 million m^3 in Tianjin, and 526 million m^3 in Hebei, respectively. The total water supply of Beijing municipality was 3.93 billion m^3 in 2018, and the reclaimed water shared about 27.4% in the total water use, much more advanced than Tianjin and Hebei province according to the 'Haihe River Basin Water Conservancy Commission of the Ministry of Water Resources' in 2019.

With the development of reclaimed water recycling, Beijing has issued relevant standards for the recycling and utilization of urban wastewater, and leadingly established a well-designed standard system for wastewater reclamation. The establishment of the system has played an effective role in promoting the reuse, recycle, and sustainable development of reclaimed water. Beijing Local Standard 'Guidelines for Agricultural Irrigation of Reclaimed Water (DB11/T 740-2010)' stipulates the basic principles, requirements, and methods for the irrigation planning and design of reclaimed water use in agriculture, as well as the monitoring and management of reclaimed water irrigation areas. According to the relevant standards, a monitoring and evaluation system for the qualities of reclaimed water, soil, drainage water, groundwater, and agricultural product are established. Forty-one newly reclaimed water plants were built during the 12th Five-Year Plan period (2011–2015), and the annual utilization of reclaimed water has increased to 950 million m^3. During the 13th Five-Year Plan period, 18 reclaimed water plants will be built, eight wastewater treatment plants will be upgraded, and 1.2 billion m^3 of high-quality reclaimed water will be utilized through engineering. Reclaimed water becomes the 'second water source' for Beijing after the South-to-North Water Diversion Project.

10.1.2.2 Development and utilization of brackish water

The distribution and utilization of underground brackish water in the NCP is mainly concentrated in the eastern low-plain area of the Hebei Plain. The saline water body with an average depth of <30 m is roughly equivalent to the bottom interface of the first aquifer in this area. Saline water with salinity >5 g/L is mainly distributed in coastal plains and frontal depressions of alluvial-proluvial fans. Saline water with salinity of 3–5 g/L is distributed in areas adjacent to coastal plains and the edge of the ancient depression in front of the fan. Brackish water with salinity of 1–3 g/L is mainly distributed in the inter-paleochannel zone and is distributed in strips.

The total amount of brackish water resources in the Hebei Plain (main composition of the NCP) is about 170 billion m^3, in which brackish water of 2–5 g/L reaches 90 billion m^3. According to the analysis of the second water resources evaluation survey in the Haihe River Basin, the average groundwater resources (the sum of precipitation, infiltration, recharge, and surface water recharge) in this region with salinity >2 g/L is 2.697 billion m^3 from 1980 to 2000, mainly distributed in the southern part of Tianjin and the eastern part of Hebei Province, with a total area of 25,401 km^2. The amount of groundwater resources with a salinity of 2–3 g/L is 1.269 billion m^3 in Hebei Province, with a distribution area of 10,812 km^2. The amount of groundwater with a salinity of 3–5 g/L is 693 million m^3, with a distribution area of 3,553 km^2. The amount of groundwater resources with a degree of >5 g/L is 499 million m^3, and the distribution area is 3,778 km^2. The 1–2 g/L brackish water exploitation in the area is mostly used for agricultural irrigation to supplement the shortage of fresh water. If the brackish water is reasonably developed and utilized, about 1 billion m^3 of groundwater resources can be increased every year in the Hebei Plain. The amount of deep fresh water exploitation can be reduced, correspondingly. And a series of environmental, ecological, and hydrogeological problems caused by groundwater over-exploitation can be alleviated. The integrated irrigation technology model of mixed irrigation of brackish and fresh water and pipelines was promoted by Hebei Province in Heilonggang area. Cangzhou City, Hebei Province, is located in the coastal low plain, which has built more than 3,300 supporting wells for mixed irrigation of brackish and fresh water, which can irrigate an area of 73,000 ha. About 60 million m^3 deep underground fresh water can be saved every year, while the irrigation cost is reduced by 150–225 RMB/ha per time.

10.1.2.3 Rainwater and floodwater harvesting and utilization

The harvesting and utilization of rainwater and floodwater also can alleviate the water shortage crisis. Some engineering measures and scientific scheduling have been carried out in recent years. Water supply increases while the ecological environment of the river is also improved as forms of rainwater harvesting in mountainous areas, as well as reservoirs, and combined regulation and storage of plain rivers-pits-ponds.

Rainwater harvesting projects in mountainous areas mainly include water cellars, pools, small ponds, and dams. In the Hebei Plain, there are about 500,000 rain harvesting projects in mountainous areas, with an annual utilization of 184 million m^3. Hebei Province is the main area of rainwater harvesting projects in mountainous areas. A total of 351,000 rainwater harvesting projects have been built, including 303,000 water cellars, 14,000 pools, and 5,000 ponds and dams. The annual water storage capacity of the rainwater harvesting project reaches 80 million m^3, supplementing the agricultural irrigation area of 128,900 hm^2.

The flood season scheduling of large reservoirs in front of Taihang Mountain and Yanshan Mountain is an important measure to improve the collection and storage of rainwater and flood resources. Raising the flood limit water level of Huangbizhuang Reservoir by 1 m can increase water storage by 40 million m^3, while raising the flood limit water level of Dongwushi Reservoir by 2 m can increase water storage by 20 million m^3. Water storage capacity of large- and medium-sized reservoirs in Hebei Province was 645 million m^3 more than the same period of the previous year in 2019. In the same year, through precise scheduling of the heavy rainfall process from August 9 to 10, the Miyun Reservoir increased the water storage capacity by 34.3 million m^3 in Beijing.

In the plain area, rivers, canals, pits, and ponds are used to jointly dispatch and collect rainwater and flood resources. Beijing has built nine cascade sluice gates (rubber dams) on the Chaobai River, with a total water storage capacity of 20 million m^3. Tianjin City has used the 'South-to-North Water Diversion Project' to divert rain and floodwater from the Beisan River to Jinghai County, Jinnan District. At the same time, primary and secondary rivers, small and medium reservoirs, deep canals, and pit ponds are used for agricultural water storage. Under the circumstance of reduced water inflow and shortage of water resources in the upper Haihe River, the potential of pits and canals for storage has been fully tapped, and rainwater resources have been actively utilized. Cangzhou City has jointly dispatched 590 rivers and canals, 4,010 pits, and three depressions, with a rainwater and flood storage capacity of about 350 million m^3 since 2009. In accordance with the principles of adapting measures to local conditions, comprehensive planning, systematic governance, and good ecology, Cangzhou City had planned to spend three years to comprehensively renovate more than 10,000 pits and ponds in the rural areas since 2018, and further improve the rainwater and flood storage capacity.

10.2 Safe utilization of brackish/brackish water resources in agriculture

10.2.1 Safe irrigation of brackish/brackish water resources

The rational development and utilization of brackish water/brackish water resources for safe irrigation is of great significance to the agricultural production and sustainable utilization of water resources in the Beijing-Tianjin-Hebei region. First of all, making full use of brackish water for reasonable irrigation can save a lot of fresh water resources including underground fresh water. Secondly, the purpose of improving brackish-alkali land can be achieved by exploiting brackish water, through adjusting the water level, reducing submerged evaporation, and restraining the rise of groundwater salinity. In addition, the large-scale exploitation of brackish water can free up underground storage capacity, replenish fresh water through rainfall infiltration, increase the regulation and storage capacity of shallow geological space, and enhance the ability to prevent waterlogging. At the same time, due to the exploitation of brackish water, the movement of groundwater is accelerated, which promotes the desalination of water, increases the amount of fresh water resources, and makes the shallow groundwater in a virtuous cycle.

In recent years, the research on brackish/brackish water irrigation technology developed rapidly. At present, there are three main ways of agricultural utilization of brackish/brackish water: direct irrigation of brackish water, mixed irrigation of brackish and fresh water, and rotating irrigation of brackish and fresh water. (1) Direct irrigation with brackish water is limited to applications when there is no fresh water resource or when fresh water resources are very scarce. In order to prevent the accumulation of salt on the surface and damage to the root system of crops, the use of this irrigation method should generally apply the advanced irrigation such as drip irrigation. Drip irrigation frequently supplies water from the ground to the soil in the distribution range of the crop root system in the form of a point water source. Part of the salt moves downward with water to achieve the purpose of leaching, and part of the salt moves to the soil around the root system with the movement of water. The small moist area around the water dropper has some salt accumulation due to the evaporation of the soil surface, so the soil in the root growth area has less salt accumulation, which is conducive to the development and growth of the root system. Drip irrigation provides a good water and saline environment for crop growth, which is the best irrigation technology for the utilization of brackish water. (2) Brackish and fresh water mixed irrigation is to use a reasonable ratio of brackish water and fresh water for irrigation or use a brackish and fresh water double drip irrigation system for water supply, in order to reduce the salt content of irrigation water or change its salt composition. The mixed irrigation of brackish and fresh water can increase the total amount of irrigated water while improving the quality of irrigation water. (3) Brackish and fresh water rotation irrigation is to use brackish water and fresh water to irrigate in turn during the crop growth period. Rotational irrigation time and water volume vary with the salinity of water, crop planting methods, and water supply conditions. For example, brackish water is used for irrigation in the salt-tolerant growth stage of crops, and fresh water is used for irrigation in the non-salt-tolerant stage of crops (such as seeding stage). In short, the key to brackish water irrigation technology is to grasp the relationship between meeting the water demand of crops and controlling the damage of salinity.

Various countries in the world have adopted different brackish water irrigation technology models according to local conditions in production practice. For example, the method of mixing deep fresh water and shallow brackish water was commonly used to reduce the salinity of brackish water for ground irrigation of field crops in Hebei Province of China. Compared with fresh water irrigation, the yield, quality, and soil salinity of cotton, wheat, and other grain crops did not change significantly by controlling the saline content of the mixed water to about 2 g/L. Research by Wang et al. (2016) showed that direct use of brackish water with salinity <3 g/L for irrigation did not show significant yield reduction of spring maize for three consecutive years. In general, the technical modes of brackish water irrigation are divided into three categories, including brackish water direct irrigation mode, brackish and fresh water mixed irrigation mode, and brackish and fresh water rotation irrigation mode. Brackish water direct irrigation mode is mainly used in serious fresh water shortage areas with good land permeability, and salt-tolerant plants are selected for cultivation (Leogrande et al., 2016). Brackish and fresh water mixed irrigation mode is to mix fresh and brackish water, and irrigate by diluting the saltwater. Brackish and fresh water rotation irrigation mode can be used on salt-tolerant and drought-resistant crops (Liu et al., 2016). In order to select an appropriate irrigation mode, crop quality, water quality, soil type, meteorological conditions, groundwater depth should be fully considered, combined with irrigation methods such as ground irrigation and drip irrigation, the corresponding agronomic measures, and the amount of irrigation and the number of irrigation times. Brackish water irrigation areas generally have good drainage conditions; combined with appropriate irrigation systems and management models, they can effectively control the accumulation of soil salinity in the root layer. Soil-soluble sodium percentage <65% and sodium adsorption ratio ≤10 is suitable for irrigation with brackish water (DB13/T 1280-2010. Technical manual for growing winter wheat with slightly brackish water irrigation). Brackish water irrigation suitability zoning should be categorized according to the climate type, brackish water quality, groundwater depth condition, and soil texture type of the irrigation area.

10.2.2 Technical case study of safe utilization of brackish water in the NCP

There are series of problems in brackish water utilization for agricultural irrigation, such as the shortage of underground fresh water resources in the NCP, the low utilization rate of brackish water in the brackish water area, the unstable mixed water quality of traditional brackish and fresh water mixed irrigation equipment, and the lack of an irrigation system for the utilization of brackish water. Therefore, it is necessary to integrate surface water and brackish water in the research and development of irrigation technology. Safety use of shallow brackish water for irrigation and reduction in the exploitation of deep water are important for improving ecological environment, and the quality and efficiency of agricultural production. This approach is suitable for wheat and maize cultivation areas in the NCP, where groundwater is over-exploited or limited, the surface water availability is not high, and the salinity of shallow brackish water does not exceed 5 g/L (Wan et al., 2012).

The key technical points of mixed irrigation technology of fresh and brackish water for winter wheat and summer maize are as follows (Jiao et al., 2021):

1. Intelligent measurement and control system: The traditional brackish and fresh water mixed irrigation equipment can only adjust and mix the brackish and fresh water according to a certain proportion. Due to the uncertainty of the water quality of the brackish and fresh water, it is difficult to reach the expected accuracy for the mixed water. Pressure transmitters and conductivity transmitters are installed in the pressure tank at the connection of shallow brackish and surface fresh water pipelines in the intelligent measurement and control system, which can monitor the salinity and pressure of the water. If the pressure is not enough, the fresh water supply will be increased; conversely if the salinity is insufficient, the salt water supply will be increased. Mixing solves the problem of unstable mixed water quality in traditional saltwater and fresh water mixed irrigation equipment. The salt and fresh water are automatically mixed in the pressure tank by the flow rate. This system supplies stable mixed water quality, which is difficult for the traditional mixed irrigation equipment.
2. Field engineering mode: For the ground irrigation engineering mode of pipeline water delivery, it is recommended to use DN125 mm main pipe and DN110 mm branch pipe. The distance between branch pipes is 48 m, and the distance between water supply hydrants is 36–48 m. The water supply hydrant is anti-theft, anti-aging, and retractable, and can freely rotate 360 degrees. For the sprinkler irrigation project mode, fully fixed PVC buried pipe or mobile PE pipe with nominal pressure of 0.63 MPa and DN75–DN125 mm for main and branch pipes can be adopted. The nozzle is in the form of a square combination, and the design spacing of the nozzle combination is a = b = 12–18 m with the height of 1.2 m.
3. The salinity of the mixed water of fresh and brackish water: The salinity of shallow brackish water is ≤5 g/L, the salinity of the mixed water of surface water and brackish water is ≤3 g/L, and the salinity of the mixed water of surface water and brackish water sprinkler irrigation is ≤2 g/L.
4. Fresh and brackish water mixed irrigation plan: The single irrigation amount of mixed surface water and brackish water ground irrigation is 825 m^3/ha, and the single irrigation amount of surface water and brackish water mixed sprinkling irrigation is 675 m^3/ha. The irrigation plan for different hydrological years is shown in Table 10.1.

10.3 Rainwater harvesting and utilization

10.3.1 Enhancing precipitation utilization through pit and pond management

The spatial and temporal distribution of rainfall in the NCP is extremely uneven. Seventy percent to 80% of the rainfall is concentrated in the three months from July to September during the flood season, and a large amount of water resources appear in the form of rain and flood. Wisely management and storage of rainfall and flood resources in the NCP can not only reduce the consumption intensity of groundwater in agricultural production, but also stabilize grain

Table 10.1 Irrigation plan of mixed irrigation with brackish and fresh water in the whole growing period of wheat and maize continuous cropping

Hydrological year	Crops	Irrigation times	Key periods of irrigation	Sprinkle irrigation (m^3/ha)	Ground irrigation with fresh water (m^3/ha)
Wet year	Wheat	2	Jointing, booting	1,350	110
	Maize	1	Soil moisture before sowing	675	55
Normal year	Wheat	3	Soil moisture before sowing, jointing, and booting	2,025	165
	Maize	1	Soil moisture before sowing	675	55
Dry year	Wheat	4	Soil moisture before sowing (or standing) jointing, booting, and milking	2,700	220
	Maize	2	Seeding and jointing	1,350	110

production capacity. This approach is beneficial for meeting the water resource demands and ensuring efficient utilization for agricultural production.

The eastern low-plain area is rich in drainage and surface water resources in the flood season. Through the joint dispatch of various water conservancy projects, the water storage capacity of the pits and ponds in this area will be fully exploited. Enough utilization of rainwater and flood resources in the flood season can scientifically control and block the transit floods and speed up the transformation of rain and flood resources into pits and ponds. Improving the storage capacity, water quality, and irrigation support capacity of pits and ponds can help stabilize the scale of regional agricultural production and improve the efficient utilization of rainwater resources.

The main pit types include pit water irrigation, abandoned kiln pits, and abandoned fish ponds. Let us take Cangzhou City in the low-plain area as an example. Among the pits and ponds surveyed (Figure 10.1), 42.9% of the pits and ponds relate to the river system, and 28.5% of them have fixed irrigation and drainage facilities, and 21.4% of them have been renovated through regular works. However, most irrigation of the rainwater and floodwater collected in the pits and ponds are carried out utilizing diesel engine-driven water pumps, which are inefficient and can only irrigate the adjacent plots of the pits and ponds.

The function of ponds to collect, divert, and irrigate water is affected by many limiting factors. First is the spatial location: (1) the location between the pit and the irrigated farmland; (2) the location and connection with the diversion ditches; (3) the location with the rural residential land (hardened ground area); (4) location and connectivity with neighboring ponds. The second factor is water storage capacity and water quality impact: (1) the area and depth of the pit and other factors that affect the overall water storage capacity of the pit; (2) the relationship with the groundwater level determines the exchange characteristics between the pit and the shallow groundwater, and affects its saline content; (3) the infiltration-groundwater exchange characteristics of the base of the pit and pond affect the seepage intensity of the rainwater/passenger water collected by the pit and pond. The third factor is the annual change

Figure 10.1 Field survey photos of rainwater storage facilities in Cangzhou Prefecture, Hebei Province. (Photograph by the author.)

dynamics of water storage in pits: (1) change in characteristics of annual water volume and water quality of pits and ponds that are mainly from rainwater harvesting and storage, and their impact on irrigation; (2) the impact of governance and engineering measures, and village drainage projects on changes in annual water volume and water quality of pits and ponds. The above constraints determine the differences in the governance methods of different pits and irrigation water patterns. Different treatment and water use schemes should be designed according to the different characteristics of pits and ponds.

The construction of pond remediation technology according to local conditions and the improvement of rainwater storage capacity are the basis of the realization of irrigation use of rainwater resources. On the one hand, it is necessary to rely on systematic and standardized treatment technology to improve the construction and management system of pit ponds. On the other hand, it is essential to continuously enhance the regulation mechanism of the water system network, and adjust and store rainwater resources on the network, as well as ensure the efficient and sustainable utilization of pits and ponds. The crops should be irrigated according to the changes of water quantity and quality in different pits and ponds during the year, plus the water demand and saline tolerance of crops, combined with different irrigation methods.

In accordance with the principles of adapting measures to local conditions, comprehensive planning, systematic governance, and good ecology, Cangzhou had planned to spend three years to comprehensively rectify and upgrade more than 10,000 pits and ponds in the rural areas since 2018. Through the construction of pits and ponds, the capacity of water storage and drainage has been improved, groundwater exploitation has been reduced, irrigation costs have been lowered, and the ecological environment has been enhanced. Take Dongguang County as an example. It is a serious resource-based water shortage area, and its per capita water resources is one-ninth of the national average. Through the implementation of the 'one village, one pit and pond' project, 447 villages in the county have 398 pits and ponds with function of water storage and irrigation, the total area of which reached 657 ha. The water storage capacity of these pits and ponds is nearly 30 million m^3 at one time, and the controlled irrigation area reaches 40,000 ha. It means that water storage fully guarantees the agricultural irrigation needs of more than 80% of the agricultural arable land in the county.

10.3.2 Integrated rainwater harvesting and utilization technologies

Rainwater harvesting and utilization is a technology integration that maximizes the benefits of rainwater achieved by adjusting the collection and storage of engineering facilities, and efficient water-saving irrigation technology. It includes rainwater harvesting, purification, storage, and water-saving irrigation technologies. The influencing factors of rainwater harvesting and utilization mainly include rainfall, runoff, topography, soil, vegetation, human factors, and rainwater utilization. Whether rainwater can be collected is closely related to the temporal and spatial distribution of rainfall, including rainfall period, amount, intensity, and frequency. The second is the land policy factor. According to the current requirements of the land authority, ditches larger than 2 m in the cultivated land will lead to change in the attributes, hence, they will not agree to build large water storage facilities. The construction of rainwater harvesting and utilization projects requires a certain amount of investment. Input-output benefit is not obvious in areas where water shortage is not serious, which will also limit the promotion and application of the technology.

The shallow groundwater and surface water in the central and southern part of Tianjin with high salinity are difficult to meet the water quality requirements for vegetable planting, and the extraction of deep groundwater is restricted. By innovative rainwater harvesting and utilization technologies, such as using greenhouse roofs, integrating shed and cellar rainwater storage, and implementing water and fertilizer integration technologies, significant advancements have been made. Using rainwater harvesting to replace extracting the groundwater enables the cultivation of rain-fed or semi-rain-fed vegetables. This integrated technology provides a model for farmers to increase production and efficiency. This technology has the effects of saving water, fertilizer, and labor, improving water quality, and increasing production and efficiency.

10.4 Wastewater irrigation

Wastewater, considered a supplemental alternative water resource, offers an effective solution to alleviate fresh water scarcity in agricultural irrigation. According to the UN Food and Agriculture Organization (FAO) in 2010, untreated or partially treated wastewater is applied to more than 20 million ha of land worldwide.

Compared to clean water, wastewater irrigation presents both advantages and drawbacks. Its main advantages lie in the stability of flow, especially in domestic and industrial wastewater, which remains consistent regardless of seasonal and climatic variations, making it readily available for irrigation after appropriate purification treatment. Moreover, wastewater contains nutrients that can boost crop growth and reduce chemical fertilizer use. Compared with clean water irrigation, wastewater irrigation can save up to 45% of fertilizer applied for wheat and the ratio can reach 94% for alfalfa (Balkhair et al., 2013). However, the limitations of wastewater irrigation are much more evident. Prior to use, wastewater must undergo complex pretreatment processes. Improperly purified wastewater used for irrigation may lead to the accumulation of heavy metals in crops and soil, resulting in declines in both agricultural product quality and soil fertility, and even contamination of shallow groundwater (Gola et al., 2016; Teklehaimanot et al., 2015). Therefore, it is essential to establish more sophisticated wastewater treatment systems and implement stringent regulatory frameworks to ensure balanced management of wastewater irrigation, aiming to minimize the adverse environmental impacts. With social and economic development, wastewater irrigation would be an essential supplementary water resource to ensure sustainable agricultural practices.

10.5 Summary

Brackish/saline water, rainwater/floodwater, and reclaimed water are the three main types of abnormal water, which is an important alternative source for agricultural irrigation. The history of abnormal water irrigation can be traced back to the 1950s in China, and has now reached an irrigation area of 3 million ha. Especially in the NCP, where water resource is scarce but the demand for food production is urgent, abnormal water plays a critical role in supplementing irrigation water and ensuring crop yields. Consequently, more attention should - be paid to the safe utilization of abnormal water. The Chinese government has enacted several regulations on abnormal water reuse quality to minimize the risks on crops and the environment, and multiple wastewater harvesting and utilization technologies have also been developed and promoted. Further scientific research on the agricultural reuse of abnormal water is necessary for data accumulation and for the sustainable development of abnormal water irrigation systems.

References

Balkhair K S, El-Nakhlawi F S, Ismail S M, Al-Solimani S G. 2013. Treated wastewater use and its effect on water conservation, vegetative yeild, yield components and water use efficiency of some vegetable crops grown under two different irrigation systems in Western Region, Saudi Arabia. *The 1st Annual International Interdisciplinary Conference*, Azores, Portugal, pp. 395–402.

Bixio D, Thoeye C, Koning De J, Joksimovic D, Savic D, Wintgens T, Melin T. 2006. Wastewater reuse in Europe. *Desalination*, 187(1–3): 89–101.

Gola D, Malik A, Shaikh Z A, Sreekrishnan T R. 2016. Impact of heavy metal containing wastewater on agricultural soil and produce: relevance of biological treatment. *Environmental Processes*, 3: 1063–1080.

Jiao Y, Wang H, Zhang S, Chen W, Zheng C. 2021. Effects of sprinkling irrigation with brackish and fresh water mixing on yield of wheat and maize and movement of soil water and salt. *Agricultural Research in the Arid Areas*, 39(06): 87–94 (in Chinese).

Leogrande R, Vitti C, Lopedota O, Ventrella D, Montemurro F. 2016. Effects of irrigation volume and saline water on maize yield and soil in southern Italy. *Irrigation and Drainage*, 65(3): 243–253.

Liu X, Til F, Chen S, Shao L, Sun H, Zhang X. 2016. Effects of saline irrigation on soil salt accumulation and grain yield in the winter wheat-summer maize double cropping system in the low plain of North China. *Journal of Integrative Agriculture*, 15(12): 2886–2898.

Lundy L, Revitt M, Ellis B. 2018. An impact assessment for urban stormwater use. *Environmental Science and Pollution Research*, 25(20): 19259–19270.

Ngigi S N, Savenije H H G, Thome J N, Rockström J, de Vries F W T P. 2005. Agro-hydrological evaluation of on-farm rainwater storage systems for supplemental irrigation in Laikipia district, Kenya. *Agricultural Water Management*, 73(1): 21–41.

Teklehaimanot G Z, Kamika I, Coetzee M A A, Momba M N B. 2015. Seasonal variation of nutrient loads in treated wastewater effluents and receiving water bodies in Sedibeng and Soshanguve, South Africa. *Environmental Monitoring and Assessment*, 187(9): 1–15.

Wan S, Jiao Y, Kang Y, Hu W, Jiang S, Tan S, Tan J, Wei L. 2012. Drip irrigation of waxy corn (Zea mays L. var. ceratina Kulesh) for production in highly saline conditions. *Agricultural Water Management*, 104: 210–220.

Wang Q, Huo Z, Zhang L, Wang J, Zhao Y. 2016. Impact of saline water irrigation on water use efficiency and soil salt accumulation for spring maize in arid regions of China. *Agricultural Water Management*, 163: 125–138.

Chapter 11

Innovation of policy system and smart irrigation technologies for a better groundwater governance

Hongjun Li, Yongqing Qi, Tao Quan, Yanjun Shen, and Dengpan Xiao

11.1 Introduction

With the growth of world's population, the increasing demand for water in agriculture to ensure food security has led to agriculture becoming the largest user of water resources, consuming 70% of global freshwater resources (Bruinsma, 2017), which further exacerbates the water crisis, especially in arid and semi-arid regions where the shortage of water resources affects local sustainable development. Agricultural irrigation has a long history in the North China Plain (NCP), and is an important practice to protect crops against drought and promote grain yield. Over the last four decades, large-scale exploitation of groundwater for irrigation not only dramatically improved grain yield and agricultural productivity, but also has led to great depletion of groundwater, triggering a series of ecological and environmental geological problems and threatening the sustainable development.

Faced with the challenge of water scarcity and excessive consumption of groundwater, the government has begun to implement very strict systems to manage the limited water resources of the country or region, and the policies of water bureaus are usually implemented in a top-down manner. The top-down water resources management policies have a significant impact on agricultural irrigation and farmers' decisions. The "Three Red Lines" policy, which set clear and binding restrictions on the use, efficiency, and quality of water resources (Chen, 2009), became the fundamental principle of water resources governance policies. In early 2012, the State Council announced that the policy would limit the country's total water use to less than 700 billion m³ per year in 2030. In 2014, in order to solve the problem of groundwater overexploitation in the NCP, the Ministry of Water Resources, the Ministry of Finance, the National Development and Reform Commission, and the Ministry of Agriculture and Rural Affairs of China jointly issued the *Action Plan for Comprehensive Management of Groundwater Overexploitation in the North China Plan*, focusing on the management measures of "saving", "control", "regulation", and "management". "Saving" means to further tap the potential of water saving, improve the efficiency and effectiveness of water resources utilization. "Control" and "regulation" mean limiting pumping and regulating planting structures, respectively. "Management" means implementing the strictest water resources management system and strengthening the construction of groundwater monitoring capacities.

Moreover, the government began to increase the introduction and promotion of water-saving irrigation technologies to further improve the level of agricultural water saving. Irrigation of crops relies on high-standard pipeline water infrastructure, supporting small bed irrigation measures. The large-scale management of farmland and cash crop planting area vigorously promotes sprinkler irrigation and micro-irrigation engineering technology. In the surface water irrigation areas, emphasis should be placed on the promotion of canal impermeability and water-saving irrigation projects. In terms of water resources management, water extraction is mainly measured by the monitoring of the groundwater level, and the construction density of monitoring wells reaches 4.16 wells/100 km^2 (Liu, 2017). In recent years, some typical cases in the NCP provide good examples for analyzing the gains and losses of water resources management policies (Wang et al., 2016).

From the perspective of improving water resources utilization efficiency, precision and intelligent irrigation is the key to implement efficient water resources management, and soil moisture monitoring and prediction are prerequisites for precision irrigation. In agriculture, soil moisture in the root zone is the only water source available to crops, and irrigation is needed to keep the soil moist to maintain high crop yields. In field water management and irrigation decision-making, evapotranspiration is the main way of soil water consumption, and soil water status determines the timing and amount of irrigation. Monitoring the status and dynamics of soil water in farmland can provide theoretical and practical guidance for irrigation. At present, there are two main methods for monitoring soil moisture: on-site monitoring at field scale and satellite remote sensing monitoring at regional scale (Qin et al., 2021). The Internet of Things (IoT) technology has greatly improved the automation and intelligence level of soil water content monitoring and irrigation control, and has become a key research and promotion technology in the field of agricultural water management, with good application prospects.

For the large number of agricultural wells without metering facilities, the method of "electricity to water" can be used to measure the water withdrawal indirectly, effectively solving the metering problem of well water pumpage. Agricultural electricity consumption can be converted into water quantity by multiplying it by the electric-water conversion coefficient. Although this method is economical and practical, it requires accurate measurement of the conversion coefficient between electricity and water. In addition, electricity metering has put forward the upgrading requirements of traditional power grid technology. Modern advanced sensor measurement technology, communication technology, information technology, computer technology, control technology are highly integrated with the IoT-based power grid, forming a new type of smart grid and generating electricity consumption big data. Mining of agricultural irrigation electricity consumption data can address the spatial heterogeneity of agricultural water use at a regional scale, while providing a technical method for implementing electricity-based agricultural water control measures.

In addition, in accordance with the requirement of "one pump one meter and one household one card", IC card intelligent water meters are installed on the pumps. Users need to first recharge and pay the water fees before they can swipe the card to pump water for irrigation. However, based on the research, we found that many upgrades to irrigation system did not produce water savings. While improvements in irrigation equipment have reduced water extraction by reducing water delivery losses, water consumption in farmland remains stable or even increase (Perry and Steduto, 2017). There are more than 16.5 million irrigation wells in the NCP, but only about 18% of them can be measured. If all wells are fitted with water meters, it will require huge financial investment from construction to post-maintenance, making it difficult to achieve greater coverage. There are many problems in the supervision and management of groundwater exploitation for agricultural irrigation, such as large land area, a large number of irrigation wells, and difficulties in the execution of water drawing permits. The statistical data of water consumption is estimated through surveys and has low accuracy, which cannot meet the requirements of groundwater monitoring and fine management. It is necessary to innovate the monitoring and management methods for groundwater exploitation.

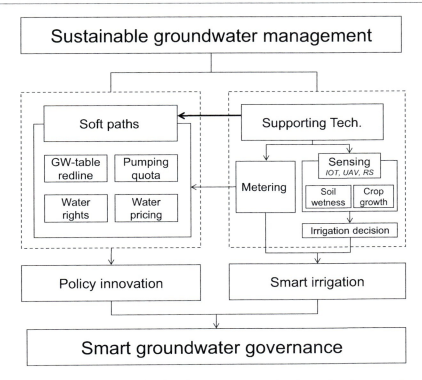

Figure 11.1 Policy innovation and related supporting technology development for effective groundwater governance.

This chapter introduces some efforts made in policy innovation and related supporting technology development to achieve a more efficient and effective groundwater management system (Figure 11.1). Following the introductions of the trials and experiences of some new policy measures such as water pricing and water rights trading that emerged in the NCP, we will also present the researches in the development of sensors and decision-making, such as monitoring soil moisture using IoT, satellite remote sensing, simulation and forecasting crop water requirements, and the irrigation decision supporting systems. Besides, some new metering technologies are also studied to help the new policies being implemented more smoothly and effectively to achieve smart groundwater governance.

11.2 Policy options of agricultural water authority and overdraft control in the NCP

11.2.1 Groundwater management policies over the past few decades

Relative policies about groundwater management and pumping control at the national and provincial levels were issued during past several decades. These policies include drilling permit requirements, pumping quota management, water charges, irrigation water price setting, and water rights system, etc.

11.2.1.1 Permit policy for well drilling

In the 1990s, some provinces in northern China began to implement formal or informal well-drilling permit policies to control the drilling of tube wells, thereby limiting the use of groundwater. Although the policy is mainly implemented at the provincial level, it is not intended to become a national policy. According to a survey of six provinces in northern China, 18% of villages had a well-drilling permit policy in 1995, and this proportion increased to 54% by 2015 (Wang et al., 2020a).

11.2.1.2 Quota management

Water quota management policy was first introduced in China in the *"National Water Law"* of 2002. It became a priority policy after the introduction of the "Three Red Lines" policy in 2012. Water resources authorities at all levels are required to determine water quotas for various water users at province, prefecture, county, irrigation district, and village levels. Under the water quota system, all water users should obtain water withdrawal permission from their superior water administrations, and the water consumption should not exceed the approved quota. In recent years, under the framework of agricultural water price reform, the concept of water resource quota has been used for groundwater pumping control. However, in using agricultural water, it is still difficult to implement water quota management due to the difficulty in calculating the generally applicable water quota and the lack of groundwater metering facilities.

11.2.1.3 Water resources fee and tax

In the *National Water Law* of 2002, water fee policy was introduced. Water resources fee can motivate people to use groundwater more efficiently (Kemper, 2007). This is especially true if this fee is linked to groundwater usage. In 2006, the Regulations on the Management of Water Abstraction Permit and Collection of Water Resources Fee were promulgated. The fee is collected by the water administration departments at the county level in accordance with the approved water withdrawal permits. It is based on actual water withdrawals and charging standards (Shen, 2015). There is an ongoing transformation from a water resources fee to a water resources tax to enhance its potential role. In 2016, Hebei Province was selected as the pilot province to shift water fee to tax and obtained significant positive effects on promoting water saving in domestic and industrial sectors. Since 2017, the "fee to tax reform" has been expanded to nine provinces in northern and western China. But the fact that millions of irrigation wells lack of metering facilities put a huge obstacle to collecting water fees or taxes from individual farmers.

11.2.1.4 Irrigation water price policy

Over the past 40 years, the reform of China's irrigation water-pricing policy has made some progress, which is mainly related to the price of surface water resources for irrigation, and not related to the groundwater. Generally, the major investors in groundwater irrigation are farmers or village collectives. They mainly pay for electricity or diesel fuel used for pumping water, and do not need to pay for the water resources fee. The collection of groundwater resource fee is the expectation of the government for the reform of irrigation water price. However, due to the high implementation cost of collecting such fees, this expectation has not been realized so far.

Due to the lack of metering facilities and high implementation costs, it is difficult to implement groundwater irrigation fee policy based on water usage in the field. Since the main operating cost of tube wells is the electricity, irrigation water fees are often charged based on electricity usage. Charging groundwater irrigation fees according to electricity consumption can be used as an alternative to the way according to irrigation volume (Li et al., 2022; Wang et al., 2020b).

11.2.1.5 Water rights and water market system

Over the past two decades, the government has been issuing regulations to promote the development of the water rights system. The first two important regulations were issued in 2005, including "*Several Opinions on Water Rights Transfer*" and "*Establishing the Framework of Water Rights System*". In 2014, the government launched formal pilot projects in seven provinces, further accelerating the development of water rights system. However, it is more difficult to establish a water rights system to promote rights transfer among irrigation water users. In surface water irrigation areas, many farmers are unaware of their water rights, let alone the fact that the water rights can be traded. In fact, due to poor implementation and high monitoring costs, water rights certificates have no sustainable effect in reducing irrigation. In the NCP, although the water rights system has been successfully piloted at the county level, the regional agricultural water rights management and water market system have not yet been established.

11.2.2 Some cases of groundwater governing reform in Hebei Province

After 2014, under the framework of the *Action Plan for Comprehensive Management of Groundwater Overdraft in the North China Plain*, different governmental departments in Hebei Province have taken corresponding measures to achieve the goals of controlling groundwater depletion. Various pilot projects for agricultural water management have been carried out at the county level. The water governance reform has achieved some success in agricultural water metering, water rights and market systems, water prices and subsidies, and so on.

11.2.2.1 Direct measuring, water rights, and market system in Cheng'an County

As a pilot area for directly measuring groundwater consumption in agricultural irrigation, more than 3000 irrigation wells in Cheng'an County have been equipped with intelligent measuring systems. In the system, smart IC cards are used to measure and charge the irrigation electricity and water resources fees (Figure 11.2).

With the support of the direct water metering system, a county-level agricultural water rights and market system was established in Cheng'an County. The idea of this system is characterized as "pricing the water rights, increase price if exceeds the quota, and make money by selling the water rights if saved". The local government and water use authority (WUA) jointly determine the water rights quota for farmers. If the farmer's water consumption exceeds the allocated quota, he/she will be charged an additional fee of 0.1 CNY/m^3 for the exceeded amount. Meanwhile, farmers can sell their unused or saved water rights within the quota through WUA with a higher price. If the water rights saved by farmers cannot be sold out, the government will purchase all of them by a price of 0.2 CNY/m^3. This practice can motivate farmers to save water and ensure their profits (Cui et al., 2019).

Figure 11.2 Intelligent measuring system for irrigation wells in Cheng'an County. (Photograph by the author.)

11.2.2.2 "Increase price then reward and subsidy" as Taocheng's Experience

Another typical case of irrigation water price reform is the so-called "Increase price then reward and subsidy" model in Taocheng District, Hengshui City. Since 2005, this reform has been supported by the "Water Saving Society" project and managed by the government of Taocheng District.

More than 90% of irrigation in the region relies on groundwater resources. Therefore, the focus of the reform is on groundwater irrigation prices. In order to make farmers accept and support the reform, the government promises to achieve a win-win goal of reducing irrigation water while maintaining or even increasing farmers' income while saving water (Figure 11.3). All farmers are required to pay groundwater fees monthly in advance. The water fee is collected by the well managers from village, and the price has been increased from 0.65 to 0.95 CNY/kWh according to their usage of electricity in pumping. At the same time of raising water fee, the government also provide an additional subsidy of 0.15 CNY/kWh. The increased price and the governmental subsidy are deposited into the account managed by the village's WUA, known

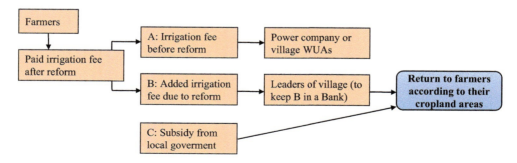

Figure 11.3 Mechanism of "increase price and provide subsidy" pilot reform for groundwater irrigation in Taocheng district.

as the "Reward and subsidy fund for water saving". Based on the total amount of water saved by farmers, as a reward, the fund is returned to farmers twice a year. The unit price calculation method for water-saving rewards is the incentive water-saving fund of the village to be divided by the total amount of water saved, compared with their total water withdrawal quota, by all the farmers of the village. Reward of the water-saving incentive fund, which is comprised by the pre-charged additional water fee from farmers (0.30 CNY/kWh) and the government subsidy of 0.15 CNY/kWh, is based on the amount of water saved by each farmer, rather than being evenly distributed to all farmers. This method can stimulate farmers' enthusiasm for water saving. Field investigations show that the amount of groundwater used by local farmers to irrigate wheat and cotton could be decreased by 21% after the reform. However, without water-saving subsidies, about half of the farmers in the region will suffer losses due to the increase in electricity/water prices. With this subsidy, most farmers in pilot villages can earn additional income through water saving (Wang et al., 2016).

From the perspective of the groundwater conservation and water-saving, the "Increase price then reward and subsidy" policy expands the price difference among farmers according to their water use efficiency, reflecting the effectiveness of the idea of "rewarding the saving user but punishing the luxury user". It not only promotes the enthusiasm of farmers to save water, but also greatly reduces the extraction of groundwater. This provides a successful case of win-win policy for solving the groundwater crisis in the NCP.

11.2.2.3 Tiered framework of water fees and indirect measurement

The tiered framework of water fees is also called the "Three lines and four ladder steps" method for determining the water fees. The proposed agricultural water-pricing framework is shown in Figure 11.4. The basic price includes the electricity fee and management fees paid to the well manager. The water resources tax is 0.1 CNY/m^3, and the water fee for water usage between water right and water limit (called price hike in the figure) is 0.1–0.2 CNY/m^3. The water-saving

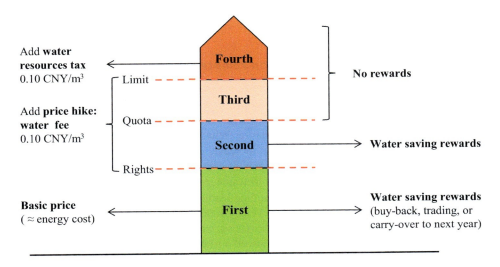

Figure 11.4 The "Three lines and four ladder steps" method for determining the water fees.

reward for grain crops staying below the water use quota is 0.2–0.3 CNY/m³. If the water use of farmer is even bellower than the water right, he/she is eligible for a water-saving reward in the form of buy-back at higher price or carry-over to the next season. There is not unified standard for the definition of the lines among counties yet. The values of the three lines depend on the local hydrogeological situation and are measured in units of m³/mu/year (15 mu = 1 ha). In Cheng'an, the values are 149.4 m³/mu/year for the water right, 222 m³/mu/year for the quota, and 296 m³/mu/year for the limit.

The method for formulating water resources tax and fee according to water usage must be supported by pumping metering facilities. Measuring millions of irrigation wells in the NCP with intelligent water meters is almost an impossible task. This not only requires a huge investment in the meters itself, but also the renovation of irrigation wells and their pipelines, and the provision of appropriate facilities to protect the meters. In addition, the maintenance cost of this system is also quite high. In Cheng'an county, the cost of a set of smart metering device is about 10,000 CNY. In 2017, Hebei Province chose the "electricity to water" approach to estimate pumping volume based on electricity consumption. Since each well is equipped an electricity meter, this method solved the measurement difficulties of a large number of irrigation wells.

11.3 Soil moisture monitoring and irrigation decision support

The traditional method of obtaining farmland information through manual investigation and on-site sampling is both time-consuming and laborious (Zheng, 2009). A solution is to use the remote sensing tools, and it is currently the only effective approach for obtaining accurate and large-scale agricultural information (Wu et al., 2019).

There are 2.67 million ha of grain sowing area and 86 large grain-producing counties in Hebei Province. A stricter groundwater management, combined with major adjustment of planting structure, is being implemented for developing a water-saving, output-stable agriculture pattern. For agricultural sustainable development in the region, it is necessary to strengthen monitoring of the regional agricultural situation. Remote sensing technology can be used for estimating agricultural water consumption or evapotranspiration, early drought warning, crop growth assessment, and crop yield estimation. Meanwhile, this information can also provide invaluable references for agricultural decision-making.

11.3.1 Drought monitoring based on remote sensing

Soil moisture content determines the magnitude of evapotranspiration, which reflects the proportion of latent and sensible heat fluxes in redistribution process of land surface energy balance. Specifically, sensible heat flux determines the variation of surface temperature, and, can indirectly serve as an indicator of soil moisture. When the soil moisture supply is sufficient, vegetation transpiration and soil evaporation increase, and radiation energy is more partitioned as latent heat. On the contrary, when soil moisture content is low, vegetation will close its leaf stomata to reduce transpiration in response to water stress. As a result, the reduction of latent heat flux means the increases of sensible heat flux, leading to an increase in surface temperature. By large-scale remote sensing monitoring, Price (1990) and Carlson et al. (1994) found that when the vegetation and soil moisture conditions varied significantly, the planar scatter plot of surface temperature and vegetation index showed a triangular relationship. That is to say, there was an surface temperature-vegetation index feature space. Based on this, some remote sensing drought indices can be established to monitor the degree and distribution of drought in a region.

11.3.2 Monitoring of regional agricultural water consumption based on remote sensing

Remote sensing not only provides surface coverage information for large areas, but also estimates the invisible water flux in reality, such as evapotranspiration. A variety of remote sensing-based evapotranspiration models, such as SEBS, TVDI, TTME, SEBAL, METRIC models, have been developed and widely applied. These models utilize the thermal infrared band, which is highly sensitive to soil moisture, to estimate surface evapotranspiration under different water stresses. In METRIC model, the surface energy balance is internally calibrated using ground-based reference evapotranspiration to reduce computational biases, and the algorithms are designed for routine application. Therefore, the METRIC model was employed and further modified to make it suitable for the evapotranspiration estimation in the NCP.

Combined with the spatial distribution of precipitation, the irrigation water demand was calculated based on water balance equation of farmland, that is, evapotranspiration minus effective precipitation. The distribution of estimated irrigation water demand is shown in Figure 11.5. The results show that in normal years such as 2016–2017 and 2018–2019, the annual irrigation water volume for winter wheat and summer corn in Hebei plain is 200–400 mm. The annual precipitation observed by meteorological stations was very high in 2017–2018, averaging amounted to 600 to 700 mm, with some areas exceeding 900 mm. Therefore, the irrigation water demand for that year is relatively small, and the negative value means that groundwater has been replenished.

11.3.3 Crop growth monitoring based on remote sensing

The high spatiotemporal resolution of remote sensing and the close correlation between sensitive bands and surface vegetation characteristics make remote sensing the main method for monitoring crop growth. The commonly used methods for monitoring crop growth include the direct monitoring method, vegetation growth process curve method, and contemporaneous comparison method. Direct monitoring method utilizes vegetation indices retrieved from remote sensing to distinguish differences in crop growth through the division of different class thresholds. MODIS NDVI data, for example, were used to detect crop growth by classifying the NDVI values into different levels. Vegetation growth process curve method uses the average value of NDVI of crop vegetation in the monitoring area by time series, to compare with the growth curves of different years in history and obtain relative information of crop growth. Contemporaneous comparison method uses the NDVI of the current crop to subtract the NDVI of last year or other reference year, and then classifies the crop growth as better, equivalent, or worse than the reference year based on the difference detected.

The change in land use during different periods can lead to the loss of comparative objects for the contemporaneous comparison method, and the difference in vegetation growth in the reference year will also lead to an overly pessimistic or optimistic assessment of the current year's growth. To overcome the drawbacks of the above methods, we propose the NDVI-based percentile method (Li et al., 2019a). This method is used to calculate the number of pixels corresponding to different NDVI values, rank the NDVI values according to their magnitudes, count the cumulative percentile of different NDVI values, and build a look-up table of NDVI percentile for the same period in the last five years. Based on the query table and the current NDVI of crops, a more objective evaluation of their growth can be made

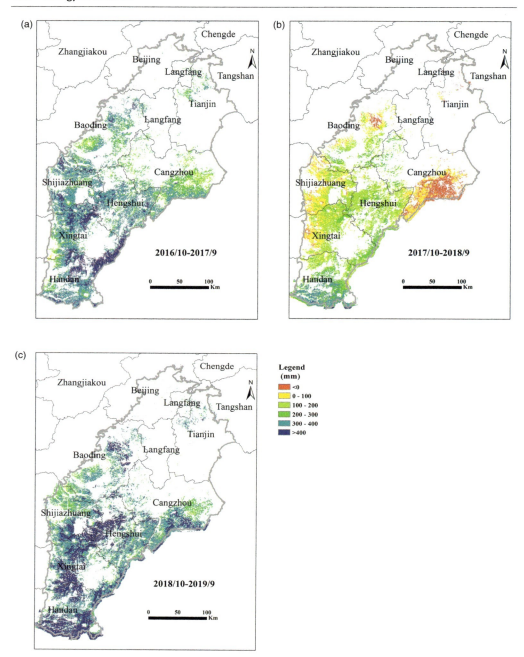

Figure 11.5 Remote sensing estimates of agricultural water deficit in the southern Hebei plain, 2016–2019. (a) Water deficit for crop year 2016/10-2017/9, (b) Water deficit for crop year 2017/10-2018/9, and (c) Water deficit for crop year 2018/10-2019/9. Negative values indicate where the annual precipitation is larger than water requirement.

11.3.4 IoT-based soil moisture monitoring technology and application

On a regional scale, the distribution of agricultural water resources (including precipitation) is often very uneven. The development of modern water-saving agriculture requires large-scale soil moisture monitoring, drought prediction, and irrigation forecasting. By which, the agriculture department can provide timely and appropriate irrigation guidance for local agricultural production, and the water conservancy administrative department can make timely decisions on water resources planning and scheduling. Therefore, it is necessary to establish an automatic monitoring network platform for agricultural conditions including soil moisture content. Based on different soil types, irrigation methods, water sources and crop types in different regions, the optimal irrigation time and water quantity can be obtained through geographic zoning combined with monitoring information. This information can be used to guide regional agricultural water management and adopt a series of water-saving measures to significantly improve the rationality of agricultural irrigation.

The combination of soil moisture monitoring technology with computer software and hardware technology, communication technology, mobile internet technology, and remote control technology has formed a modern agricultural intelligent irrigation IoT technology system. Utilizing IoT technology, we have established the Hebei IoT Agricultural Monitoring Network (http://wlw.casbhlc.com/m2) to realize real-time online monitoring of agricultural conditions and provide information services for agricultural production. In addition, combining the IoT agricultural monitoring network with remote sensing monitoring technology, we established a space-earth integrated agricultural monitoring platform.

11.3.4.1 Regional IoT agricultural monitoring network

The structure of the IoT agricultural monitoring network for Hebei Province is shown in Figure 11.6. The base of each agricultural monitoring station is built in farmland, and the required sensors are installed on the monitoring pole. Because it is in farmland, the agricultural monitoring station adopts a power supply method of solar energy and batteries. For data communication, monitoring stations use 4G wireless communication modules to communicate with cloud servers. The cloud server regularly sends data collection instructions to all monitoring stations based on the monitoring time interval setting. After receiving instructions, the monitoring station uses a data collector to collect data from all sensors on the site, packages all the data according to the communication protocol, and sends it to the cloud server through a 4G communication network. The cloud server receives, parses, and stores data in the corresponding data table according to the database format. The agricultural monitoring network system provides a data service network platform for the public, and users can view the real-time and historical agricultural monitoring information of various monitoring stations in Hebei Province through the computer or mobile phone.

11.3.4.2 Deployment and installation of agricultural monitoring stations

Agricultural monitoring sites should have good regional representativeness, that is, the monitoring data should represent the management characteristics of local agriculture, reflect crop growth, and farmland microclimatic environment. Large agricultural households or cooperatives have a larger cultivated land area, and they are relatively unified in crop types and management,

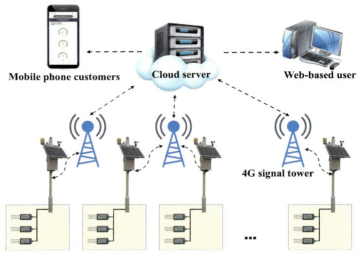

Figure 11.6 The structure of the Hebei IoT agricultural monitoring network.

so the spatial heterogeneity of the farmland is relatively little. We choose this type of farmland for setting agricultural monitoring. By screening the types of large agricultural households and cooperatives, a total of 22 sets of agricultural monitoring stations were installed in typical farmland in Hebei Province as shown in Figure 11.7.

The same sensors are installed at each monitoring station to collect the following agricultural information uniformly: soil volumetric water content (%), soil electrical conductivity (μs/cm) and soil temperature (°C) at three different depths (20, 40, and 60 cm) in the root layer of the crop, air temperature (°C), relative humidity (%), wind speed (m/s), wind direction (degree) at 3 m above ground surface, and precipitation (mm).

To monitor the crop growth intuitively, a high-resolution camera is installed at each monitoring station, and through the control of the cloud server, photos of the farmland are taken regularly and uploaded to the cloud server, providing real-time high-resolution images of the farmland crops. The installation of underground and above ground sensors at monitoring stations is shown in Figure 11.8.

The cloud server communicates with all IoT sensors through 4G modules, and sets and controls each monitoring station, including setting the collection frequency of all monitoring parameters, checking the networking and power status of monitoring stations, and recording monitoring station status logs, etc. In addition to collecting data from each monitoring station at sampling time intervals, cloud services can also obtain real-time information from each monitoring station by sending instructions.

11.3.4.3 Agricultural monitoring information service

The Hebei IoT agricultural monitoring network platform provides the monitoring information of each station to the public through the website, and users can check the current and historical crop information of each monitoring station after logging in to the website. According to the information of soil water content at three layers of crops root zone, combined with the irrigation

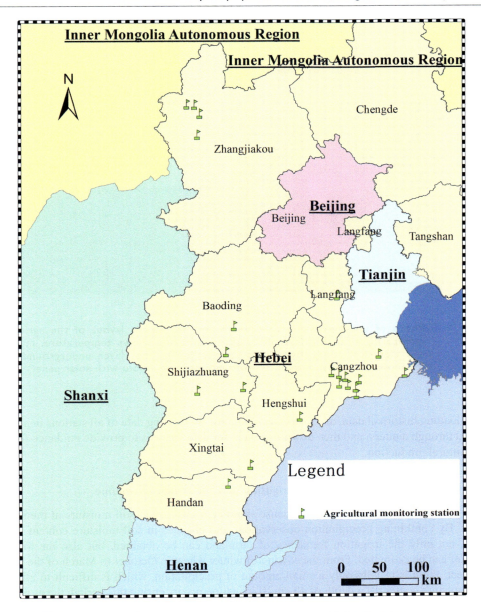

Figure 11.7 Distribution of agricultural monitoring stations in Hebei Province.

index for different crops in different regions, it can be determined whether the current crops need irrigation and irrigation amount, that is, the amount of water required for the soil moisture content of crop roots to reach their field capacity (Zhang et al., 2002). By using high-resolution images of farmland, the crop growth status of all monitoring stations in Hebei Province can be intuitively understood, thereby quickly obtaining basic information of the monitoring area. In addition, the information service website also provides users with the function of querying and

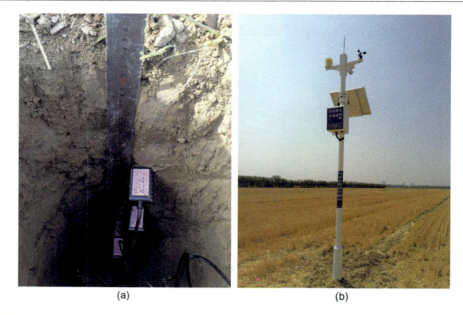

(a) (b)

Figure 11.8 Photo of underground and above-ground instrument layout of the agricultural monitoring site. (Photograph by the author.) (a) shows temperature, moisture, and electrical conductivity sensors installed in root layer underground; while (b) shows the above ground sensors mounted on a pole with solar panel.

downloading historical data, and users can obtain the monitoring data of all stations in a certain period through a query, and then download them for data analysis to provide guidance for local agricultural production.

11.3.4.4 Monitoring of farmers' irrigation and the irrigation volume

Both precipitation and irrigation can cause significant changes in soil moisture at the roots of crops. By combining precipitation observations and changes in soil moisture content of crop roots, not only the irrigation frequency of farmland can be obtained, but also the irrigation amount can be estimated. From the sowing of winter wheat in October to March of the following year, there is usually only a small amount of precipitation, which is difficult to affect the soil moisture content at a depth of 40 cm. If irrigation occurs during this period, the monitoring station will detect a sudden increase in soil moisture content at a depth of 40 cm. Winter wheat irrigation occurs more frequently from April to May. There will also be some heavy rainfall during this period. Therefore, it is necessary to couple the precipitation observations with soil moisture changes in order to detect the irrigation information. After irrigation, the increase in soil moisture content for each layer of crop roots before and after irrigation can be calculated separately. The irrigation water volume for this time can be obtained by accumulating all the increases. These analyses can reflect the irrigation habits of farmers and contribute to the improvement of water management in agriculture.

11.4 Metering pumping for precision irrigation

With the implementation of the strictest water resources management system, the monitoring of agricultural irrigation water consumption has become one of the main contents of groundwater resource control, and is also a necessary means of monitoring the total amount and utilization rate of agricultural irrigation water.

The current common measurement for agricultural irrigation water include four categories: (1) Installing mechanical or electronic water meter at the head of well. This measurement method is quite popular. However, due to the outdated technology of this water meters, manual reading and recording are required. (2) Installing an ultrasonic flowmeter at the outlet of the pump. The working principle is to detect its flow rate by receiving the reflected signal of ultrasonic waves in the fluid, and then convert it into water volume. It can realize online real-time monitoring of water extraction if combined with the wireless data acquisition system of IoT. (3) Using "water meter + IC card" for measurement and management. It can establish a management system with metering and charging through the water meter and IC card. (4) Billing management system using "electricity meters + IC card". Through this method, water pumping data can be indirectly charged and managed through electricity consumption data.

11.4.1 Irrigation water measurement using IC cards

Recently, intelligent IC card technology has been widely used in agricultural irrigation water measurement for its advantages of convenience and power saving. The IC card is used as the carrier to record the personal cost information of water users, which is safe, reliable, and not easy to damage. Generally, agricultural irrigation IC card metering and control system is composed of a central control system and an extension installed in the pump room. The central control system includes an IC card reader and an irrigation management system. The main functions of the irrigation management system mainly include: system setting, IC card management, water sales management, inquiry, and statistics.

The system setting module sets the parameters of the irrigation management system. According to the relevant regulations from the regional agricultural water resources management department, the annual irrigation quota is set for different crops, and the stepped water prices is set for different irrigation water consumption.

IC card management module is responsible for managing the information for farmers and their IC cards. Each farmer usually has multiple plots of farmland, which are irrigated using different wells. The system generally uses the way of one IC card for one plot of farmland, and links all the plots of farmland owned by farmers with the irrigation wells to which they belong through the plot number and IC card code.

The function of the water sales management module is to sell irrigation water and manage water fees. Using the water consumption quota of farmer plots as the upper limit for water sales, combined with the setting of tiered water prices, the system completes the purchase operation by reading and writing the IC card, and records relevant data.

The query and statistics module can perform queries and statistics on all information. The combination of different query conditions enables the query of information such as farmers' water purchase records, pumping quantity of each irrigation well, and IC card issuance records during different periods. The statistical function can summarize various types of data, such as regional total water consumption.

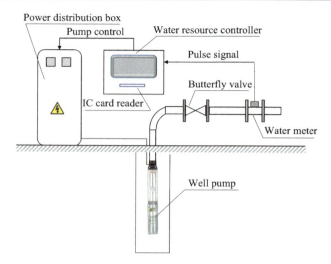

Figure 11.9 Structure of pump room equipment for agricultural irrigation IC card metering.

For the measurement of agricultural irrigation using IC cards, equipments in the pump room include a well pump, power distribution box, water resource controller, and pulse water meter. The composition structure is shown in Figure 11.9.

11.4.2 Measuring irrigation water based on the "electricity to water" method

The method of "electricity to water" is to convert electricity consumption into pumping volume by a conversion coefficient. This requires measuring both the well's electricity consumption and the amount of water being pumped. By analyzing the relationship between electricity consumption and the corresponding pumped water volume, the conversion coefficient of electricity to water is obtained. The calculation formula of the conversion coefficient of electricity to water is (Wang, 2017):

$$T_c = A_w / A_e \qquad (11.1)$$

where T_c is the conversion coefficient of electricity and water, and the unit is m³/kWh. A_w is the total amount of water pumped in m³. A_e is the total electricity consumption in kWh.

In order to standardize the agricultural water management by "electricity to water" method and promote the collection of agricultural water resources tax, Hebei Province has issued the *Detailed Rules for the Implementation of Electricity-to-Water Measurement for Agricultural Water in Hebei Province*. The rules stipulate that when measuring the conversion coefficient of electricity to water, a representative typical well should be selected for each region, and factors such as local water resources zoning, the number of underground water exploitation layers, well types, and irrigation method should be considered. According to the principle of selecting one typical irrigation well for every 80,000 mu of irrigated area, Hebei Province has selected all typical irrigation wells and ensured that they are evenly distributed.

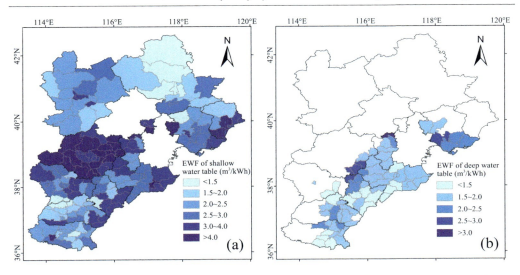

Figure 11.10 Distribution of the electric-water conversion coefficient by county in Hebei Province. (a) shallow groundwater, (b) deep groundwater table.

After actual observation and data analysis, the electricity-to-water conversion coefficients of typical shallow and deep water irrigation wells were obtained, and the average value of wells in each county was calculated as the conversion coefficient of the local area, as shown in Figure 11.10.

11.4.3 Monitoring and management of agricultural water use based on State Grid big data

At present, State Grid of China has completed the electrification of 720,000 agricultural wells in southern Hebei Province (Wang et al., 2022). This project utilizes smart meters to achieve real-time collection of agricultural irrigation electricity consumption data, constructs an irrigation electricity management system, and can provide analysis of big data. The information of farmers' irrigation water use can be obtained from the electricity consumption data by using the electric-water conversion coefficient. The big data analysis of farmer irrigation electricity can provide useful reference information for monitoring the intensity of regional groundwater exploitation and water resource management.

For the management of water resources, it is usually the responsibility of the ministry of water resources, with little involvement from other departments. Due to the fact that the big data on agricultural irrigation electricity is controlled by the State Grid, and considering the importance of electromechanical wells in groundwater irrigation, many counties have started using electricity prices as a lever to promote the rational utilization of water resources. With the help of State Grid of China, the management of groundwater exploitation can be achieved through electricity discount, water rights allocation combined with tiered water fee, and control of irrigation electricity usage. This will be a feasible and innovative way to effectively manage groundwater resources.

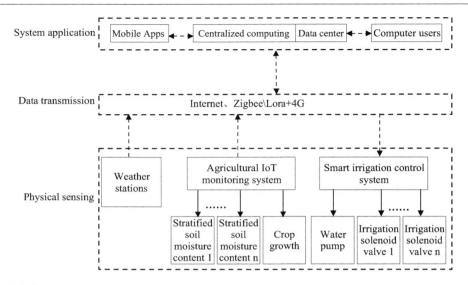

Figure 11.11 Structure of smart irrigation system based on IoT.

11.4.4 Smart irrigation system based on IoT

The smart irrigation system first needs to obtain the spatial distribution of soil moisture content, and then accurately estimate the irrigation time and water volume required for different plots based on parameters such as the current growth stage of crops, future target yields, and weather forecasts. Using irrigation schemes, the system achieves precise irrigation of different plots through remote control systems. The smart irrigation system integrates functions such as monitoring, analysis, decision-making, and implementation, which can complete the entire process of precise irrigation while considering water usage quotas, achieving a reduction in crop water consumption (Haverkort, 2007). The IoT-based soil moisture monitoring equipment solves the technical problem of rapid online monitoring of farmland moisture. In addition, these data can also provide ground synchronization support for remote sensing moisture monitoring, achieving monitoring of soil moisture in larger areas of farmland.

Usually, from the perspective of information flow, the smart irrigation system starts with soil moisture monitoring, and the information processing processes include information collection, analysis, and ultimately deciding whether irrigation is needed and the amount of irrigation, and implementing irrigation at last. Overall, the system consists of three main parts, that is, physical sensing layer, data transmission layer, and system application layer (Figure 11.11).

11.5 Summary

Agriculture is the first major water user in NCP. Achieving "zero growth" or even "negative growth" of total agricultural water consumption is the basic strategy for national water resources security in China. The support of food security under the strictest water management system requires the promotion of modern water-saving agricultural techniques. Faced with the scattered management of land under China's household contract management system and the increasing scale management after land transfer, it is necessary to adapt to the existing different production methods and management levels, and actively carry out water-saving technology innovation to improve the wisdom of agricultural water-saving management.

Smart agriculture is an indispensable and important force in the process of modern agriculture, which is also the future direction of global agricultural development. The intelligence of irrigation water management will greatly improve the utilization efficiency of agricultural water, which is of great significance in coping with the problem of water scarcity. As a modern engineering technology, water-saving irrigation must be supported by effective soil moisture monitoring and intelligent irrigation equipments, while also adapting to the agricultural development needs under water resource constraints. In this sense, the applications of information technology such as IoT, big data and cloud computing in agriculture can continuously promote the transformation and upgrading of agricultural irrigation technology. These intelligent or precision technologies will greatly support groundwater policy innovations to be implemented, although the small householder operation system makes precision policies costly and hard to be applied. The dawn of policy innovation supported by information technology has emerged in the water management practices of Cheng'an County and Taocheng County, which will also bring bright prospects for sustainable management of groundwater in the NCP.

References

Bruinsma J. 2017. *World agriculture: towards 2015/2030: an FAO study*. England, London: Routledge.

Carlson D H, Sharrow S H, Emmingham W H, Lavender D P. 1994. Plant-soil water relations in forestry and silvopastoral systems in Oregon. *Agroforestry Systems*, 25(1): 1–12.

Chen L. 2009. Implementing the strictest water management institution to ensure sustainable development of socio-economy. *China Water Conservation*, 5: 9–17.

Haverkort A J. 2007. Ancha Srinivasan (ed): handbook of precision agriculture. Principles and applications. *Euphytica*, 156(1): 269–270.

Kemper K E. 2007. Instruments and institutions for groundwater management, in Giordano M, Villholth K G. (Eds.), *The agricultural groundwater revolution: opportunities and threats to development*, pp. 153–172. Wallingford: CABI.

Li B, Chen X, Xu W. 2019a. Drought monitoring analysis in North China based on DEM correction of TVDI. *Journal of East China University of Technology*, 42(3): 266–273.

Li C, Li H, Li J, Lei Y, Li C, Manevski K, Shen Y. 2019b. Using NDVI percentiles to monitor real-time crop growth. *Computers and Electronics in Agriculture*, 162: 357–363.

Liu S. 2017. *Evaluation system building for groundwater overexploitation zone management*. Beijing, China: China Institute of Water Resources and Hydropower Research.

Perry C J, Steduto P. 2017. *Does improved irrigation technology save water? A review of the evidence*. Cairo, Egypt: Food and Agriculture Organization of the United Nations, p. 42.

Price J C. 1990. Using spatial context in satellite data to infer regional scale evapotranspiration. *IEEE Transactions on Geoscience and Remote Sensing*, 28(5): 940–948.

Qin X, Pang Z, Jiang W, Feng T, Fu J. 2021. Progress and development trend of soil moisture microwave remote sensing retrieval method. *Journal of Geo-Information Science*, 23(10): 1728–1742.

Shen D. 2015. Groundwater management in China. *Water Policy*, 17(1): 61–82.

Wang H, Li H, Qi Y, Dong Z, Li F, Yan C, Shao L, Zhang X. 2022. Development of a decision support system for irrigation management to control groundwater withdrawal. *Chinese Journal of Eco-Agriculture*, 30(1): 138–152.

Wang J. 2017. Research and application of "electricity for water" method. *China Water Resources*, (11): 34–35.

Wang J, Jiang Y, Wang H, Huang Q, Deng H. 2020a. Groundwater irrigation and management in northern China: status, trends, and challenges. *International Journal of Water Resources Development*, 36(4): 670–696.

Wang J, Zhang L, Huang J. 2016. How could we realize a win-win strategy on irrigation price policy? Evaluation of a pilot reform project in Hebei Province, China. *Journal of Hydrology*, 539: 379–391.

Wang L, Kinzelbach W, Yao H, Steiner J, Wang H. 2020b. How to meter agricultural pumping at numerous small-scale wells? An indirect monitoring method using electric energy as proxy. *Water*, 12(9): 2477.

Wu B, Zhang M, Zeng H, Yan N, Zhang X, Xing Q, Chang S. 2019. Twenty years of CropWatch: progress and prospect. *Journal of Remote Sensing*, 23(6): 1053–1063.

Zhang X, Pei D, Hu C. 2002. Index system for irrigation scheduling of winter wheat and maize in the Piedmont of Mountain Taihang. *Transactions of the CSAE*, 18(6): 36–41.

Zheng G. 2009. Scientific response to global warming & improving food security guarantee capacity. *QIUSHI*, 2009(23): 47–49.

Index

aquifer 3–4, 127, 143, 148

basin irrigation (BI) 21, 44, 83, 85
Beijing-Tianjin-Hebei (BTH) 109, 112, 127, 131, 163, 165
biomass production 71, 73, 74–76, 80, 85
blue water 19
Bo Sea 4
business as usual scenario (BAU) 136

calorie consumption 3
canal impermeability 38, 174
canal utilization efficiency 38
canopy coverage 74
canopy temperature 76
canopy transpiration 50
carbon assimilation 89
carbon exchange 57
carbon isotope discrimination 76
carbon sequestration 73
chlorophyll content 43, 78
Classification and Regression Tree method (CART) 98
climate change 4, 14, 76, 157
crop coefficient 105, 115, 123
crop evapotranspiration 104
crop rotation 54, 60, 122, 124
crop transpiration 33, 35, 71, 86

deep drainage 44, 54, 148–151
deficit irrigation 20, 22, 75, 80, 89
depth to groundwater level (DTW) 8
drip irrigation (DI) 67–69, 83, 165–166
drought event 4
drought resistance 32, 43, 74, 76
drought risk 14–15
drought tolerance 42
dry matter accumulation 59, 77–78
dry matter production 74, 79

economic water productivity ratio 85
ecosystem productivity 52, 57
electricity consumption 14, 174, 180, 187–189
energy budget 50, 52, 54

FAO penman-monteith method 104–105
fertilizer application 17, 20, 43, 52, 97–98
fertilizer management 17
field capacity 62, 87, 117, 155, 185
flood irrigation (FI) 4, 67, 83
food demand 3–4, 15, 17, 89, 97, 131–142
freshwater resources 3, 173

genetic engineering 43
genetic gain 74–75
grain filling 22, 51, 78, 79, 82, 86
groundwater depression 4, 18
groundwater extraction 14, 137, 147–148, 155, 157
groundwater governance 21, 173, 175
groundwater monitoring 24, 173, 174
groundwater over-exploitation 18–21, 112–114, 139, 143, 164
groundwater pollutants 12
groundwater pollution 157
groundwater pumpage 11–12
groundwater pumping 17, 19, 21, 46, 136–137
groundwater quality 12, 25, 147–158
groundwater recharge 14, 20, 122, 147–153

Harmonic Analysis of Time Series (HANTS) 98
harvest index (HI) 40, 43, 71
hydrogeological condition 148

Internet of Things (IoT) 21, 174–175, 183–184, 190–191
irrigated cropland 6, 70, 153
irrigation amount 14, 22, 67–69, 82–85, 185–186
irrigation application 74, 82, 83

irrigation efficiency 20, 38
irrigation forecasting 42
irrigation frequency 22, 57, 72, 83–86, 186
irrigation management 37, 42, 72, 151, 187
irrigation scheduling 69, 71, 80, 85, 192
irrigation technology 44, 144, 161–167, 170, 191
irrigation timing 72
irrigation water demand 3, 113, 135, 141, 181
irrigation water fees 177
irrigation water management 191
irrigation water supply 83
isotope method 32, 60–62

land reclamation 97, 100, 104
land subsidence 18, 26, 147
land use change 4, 13, 97–104, 160
land use patterns 110
Landsat TM/ETM 98
latent heat flux (LE) 50–51, 53–56, 180
leaf area index (LAI) 87, 115
leaf chlorophyll content 78

Moderate-resolution Imaging Spectroradiometer (MODIS) 98, 181

net ecosystem productivity (NEE) 57
net income 23, 83–86, 120
net radiation 39, 50, 53–56
nitrate availability 73
nitrate concentration 12–13, 20
nitrate leaching 12, 20, 28, 151–155
nitrate pollution 18, 25, 28, 155
nitrate supply 89
nitrogen application 151
Normalized Difference Vegetation Index (NDVI) 98, 105, 181
nutrient holding capacity 73
nutrient mass flow 73
nutrient uptake 85

organic matter 38, 73
oxidative stress 89

photosynthetic rate 38–41, 76
pillow irrigation 69, 83, 85
pipe irrigation 21
plant height 76
plant transpiration 34, 62, 88
population growth 3–4, 97, 132, 134
precipitation use efficiency 37, 40
pumping quota 24, 175

redline control 24
remote sensing (RS) 21, 98–105, 122, 174–175, 181–183

root architecture 78
root biomass 78
root chemical signals 89
root distribution 76
root hydraulic resistance 89
root length density (RLD) 76
root water uptake 32, 62–63

sensible heat flux (H) 50, 54–56, 180
soil compaction 73
soil fertility 33, 72, 88, 171
soil heat flux (G) 32, 39, 50–56, 180
soil moisture content 183–186
soil nutrient contents 73
soil organic matter 73
Soil-Plant-Atmosphere Continuum (SPAC) 31–37, 45, 50
soil porosity 43
soil salinization 44, 97, 100, 162
soil temperature 25, 52, 87–88, 184
soil texture 53, 155, 166
soil water depletion 81, 84–85, 119
soil water drainage 51
soil water storage 43, 51, 54–57, 148, 158
South-to-North Water Diversion Project (SNWDP) 11, 18, 122, 134, 155, 163–165
stomatal conductance 40–41, 89
straw mulching 24–25, 43, 87–88
straw return 73
subsurface drip irrigation (SSDI) 44, 48, 52, 68
surface drip irrigation (SDI) 42, 44, 48, 52, 67–68
surface irrigation 6, 44, 69, 85
surface mulching 42
surface runoff 51

Taihang Mountains 4, 8, 106, 112
tillage layer 73
tube-sprinkler irrigation (SI) 69, 83

vadose zone 13, 18, 20, 147–158
virtual water 19, 23, 134, 136

water and energy balance 69
water availability 82–83, 89, 127
water crisis 3, 71, 113, 173
water cycle 31–34, 50, 52, 148
water distribution 92
water flux 32, 34, 39, 60, 73, 181
water footprint 111
water infiltration 73
water loss 19, 23, 37, 44, 83, 89
water metering 14, 42, 176, 177
water pollution 23, 71, 157
water potential 31, 33
water price 16, 86, 175–179, 187

water productivity (WP) 14, 67, 69, 71, 85, 157
water resources management 44, 173–174, 187
water resources utilization 139–140, 173–174
water retention 24
water rights 14, 45, 175–177, 189
water sensitivity 71
water transfer 11, 31, 34, 45, 113

water use efficiency (WUE) 14, 20, 109, 113, 129
winter wheat-summer maize double cropping system 17, 24, 100, 104, 110, 113–114

Yan Mountains 4
Yellow River 4–10, 13, 104, 106–108
yield response 46, 117

9781032116747